Rosalind Franklin

By the same author

Beyond Babel

The Half-Parent: Living with Other People's Children

Who's Afraid of Elizabeth Taylor?

The Marrying Kind

Nora: A Biography of Nora Joyce

The Married Man: A Life of D. H. Lawrence

Yeats's Ghosts: A New Life of W. B. Yeats

Rosalind Franklin

THE DARK LADY OF DNA

BRENDA MADDOX

HarperCollins*Publishers*

HarperCollins books may be purchased for educational, business, or sales promotional use. For information, please write: Special Markets Department, HarperCollins Publishers Inc., 10 East 53rd Street, New York, NY 10022.

First published in the United Kingdom in 2002 by HarperCollins Publishers.

FIRST EDITION

Printed on acid-free paper

Library of Congress Cataloging-in-Publication Data is available upon request.

ISBN 0-06-018407-8

03 04 05 06 UK/RRD 10 9 8 7 6 5 4 3 2

For John

CONTENTS

Part Three

ILLUSTRATIONS

ILLUSTRATIONS IN TEXT

ACKNOWLEDGEMENTS

Rosalind Franklin apart, the principals in the drama of the discovery of the structure of DNA in 1953 survived into the twenty-first century. I have been fortunate to have had many discussions with Dr Francis Crick, Dr James Watson, and Professor Maurice Wilkins. All three have been generous with their time, their recollections, opinions and permissions to quote unpublished letters. Among Rosalind's close associates, those who have shown great interest and offers of assistance include Professor D.L.D. Caspar, Dr K.C. Holmes, Professor Raymond Gosling, Sir Aaron Klug and Dr Vittorio Luzzati.

As I went along, I was surprised (and relieved) by the willingness of scientists to discuss their work with someone with a rudimentary scientific vocabulary. Their patience is explained, I think, by the inherent openness of science as well as by a sincere wish to help set a tangled record straight: Dr John Finch, Professor Bruce Fraser, Dr Durward W.J. Cruickshank, Dr William Ginoza, Professor Alan Mackay, Dr Peter Pauling, Dr Margaret Nance Pierce and Professor H.R. Wilson are among those who sent material from their personal archives, as did the historian who did the research for the BBC's 1987 film, *Life Story*, Jane Callander, Dr June Goodfield and Horace Freeland Judson. That so many volunteered to help is owing to the kindness of the editor of *Nature*, Dr Philip Campbell, who published my letter announcing the planned biography. The scope of the response was a small reminder of *Nature*'s global reach and influence.

A great privilege accorded me was access to Rosalind's vivid personal letters, written from her childhood until her last weeks of life. I am deeply grateful to Jenifer Franklin Glynn, and also to Colin and Roland Franklin, for allowing liberal quotation from these without asking to authorise or even agree with the resulting book. These quotations allow Rosalind to speak in her own voice at last, and the real person to emerge from the accretion of myth and caricature.

I am also indebted to those who read the manuscript in rough form, Professor Paul Doty, Dr Walter Gratzer, Dr K.C. Holmes, Professor Ian Glynn, Sir John Maddox and Bernard McGinley. I am grateful, as in previous biographies, for the expert eye on photographs of the historian of twentieth-century fashion, Jane Mulvagh. Any errors remaining are unquestionably my own.

Dr Gunther Stent's masterly critical edition of *The Double Helix* was invaluable, with its inclusion of the three DNA papers which appeared in *Nature* on 25 April 1953, as well as the original reviews of Watson's book and subsequent articles stimulated by the controversy that followed publication.

This biography could not have been written without the resources of many libraries and archives. The principal collections of Rosalind Franklin papers outside the family holdings are in the Anne Sayre Archive held by the American Society for Microbiology Archives at the University of Maryland, Baltimore County, under the supervision of Jeff Karr, archivist; in the Franklin papers at Churchill College Library, Cambridge, which also holds the J.T. Randall papers, and in Jeremy Norman's Archive of Molecular Biology at Novato, California, whose holdings include the papers of Aaron Klug, Max Perutz and some of Rosalind herself.

Many other libraries and archives contributed documents, information and answers to queries. I would like to express my appreciation to the Bank of England (David Parr), Bexhill Library (Brian Scott), Bexhill Museum (Julian Porter), Bodleian Library, Oxford (the Charles Coulson papers), Central Library, Royal Borough of Kensington and Chelsea, Cold Spring Harbor Press (Dr John Inglis), East Sussex Record Office, Gordon Research Conferences (Barbara Henshel), IACR-Rothamsted, King's College

London (Patricia Methuen), the Medical Research Council (Tom Hudson), the National Portrait Gallery, Newnham College, Cambridge (Anne Thomson), Norland Place School (David Alexander), the Novartis (formerly Ciba) Foundation, the University of Oregon (the Ava Helen and Linus Pauling Papers, archivist Chris Petersen), the Public Record Office of the United Kingdom, the Royal Institution (Frank James), The Royal Society, the Science Museum Agricultural, St Paul's Girls' School (Howard Bailes, archivist), University College Hospital and the Working Men's College for Women and Men (Satnam Gill).

For hospitality, encouragement, and often the loan of books and papers, I am grateful to Dr and Mrs Carl Djerassi, Professor Paul and the late Helga Doty, Professor and Mrs Hubert Dreyfus, Mrs Myrtle Franklin Ellenbogen, Mr and Mrs Colin and Charlotte Franklin, Mr and Mrs Roland Franklin, Professor and Mrs Ian Glynn, Dr and Mrs Raymond Gosling, Mrs Pauline Cowan Harrison, Professor Sir Aaron Klug and Lady Klug, Dr Mair Livingstone, Dr Jacques Maire, Dr Joan Mason, Mr and Mrs Jeremy Norman, Maureen Howard and Mark Probst, Ann Satterthwaite, Dr David Sayre, Dr Gunther S. Stent, Professor and Mrs Robert Tracy.

For interviews, conversations, e-mails and helpful comments, I would like to thank: Dr Simon Altmann, Edgar Astaire, Dr Alec Bangham, John Barton, Marianne Weill Baruch, Dr Stanley Bayley, Anthony Bourne, Dr John Bradley, Dr Sydney Brenner, Baroness Brigstocke, Drs Geoffrey and Angela Brown, Sir John Cadogan, C.S. Carlson, Ruth Carr, Dr Erwin Chargaff, Dr Carolyn Cohen, Freda Ticehurst Collier, Dr Francis and Mrs Odile Crick, Professor D.W.J. Cruickshank, Dr Philip D'Arcy Hart, Professor Edward Deeley, Dr Jack Dunitz, Xavier Duval, Dr Lynne Elkin, Dr Gary Felsenfeld, Georgina Ferry, Dr John Finch, Dr Jeni Fordham, Norman Franklin, Dr Mary Fraser, Rachel Glaeser, Dr William Ginoza, Vincent Gray, Professor Dr Drago Grdeniè, Istuan Hargittai, Dr Peter Harris, Dr Pauline Cowan Harrison, Dr Louise Heller, Dr Philip Hemily, Dr Peter Hirsch, Professor Eric

Hobsbawm, Carol Howard, Barbara Izdebska, Dan Jacobson, Rosalind Franklin Jekowsky, François Xavier Keraly, Jean Kerlogue, Professor G.D.S. King, Dr W.G.P. Lamb, Stan Lenton, Dick Leonard, Sylvia Castle Levinson, Margaret Levy, Barbara Little, Dr Mair Livingstone, Warner E. Love, Jacques Maire, Marie Marcus, Dr Joan Mason, John Mason, Dr Matthew Meselson, Sir John Meurig Thomas, Mr and Mrs Jeremy Norman, Drs A.C.T. and Margaret North, Dr Agnès Oberlin, Dr Peter Pauling, Martyn Peese, the late Dr Max Perutz, Dr Margaret Nance Pierce, W.S. Pierpont, Anne Crawford Piper, Yves Pomeau, Dr and Mrs Alexander Rich, Noel Richley, Ursula Franklin Richley, Lord Samuel, Al Seckel, Dr Albert Siegel, Dr Bob Simmons, Dr Bea A. Singer, Dr Alex Stokes, Valerie Sutton-Mattock, Dr Denise Tchoubar, Dr Irwin Tessman, Dr Wolfie Traub, Peter Trent, David Turnball, Mrs James Watt, Dr Robley Williams, Dr Bryon Wilson, Professor Herbert Wilson, Dr Jan Witkowski, Evi Wolgemuth and Dr Lewis Wolpert.

My thanks must also go to Michael Fishwick at Harper Collins UK and Terry Karten, HarperCollins US, who have enthusiastically supported this project from the beginning; to Kate Johnson, the most scrupulous of editors, and to Helen Ellis and Jane Beirn, for their publicity efforts. For many years I have been fortunate to have, as agents and friends, Caradoc King at A.P. Watt in London, and Ellen Levine of the Ellen Levine Agency in New York.

Once more John, Bronwen and Bruno Maddox have sustained me through a book with patient listening and perceptive comment.

PROLOGUE

> 'It has not escaped our notice that the specific pairing we have postulated immediately suggests a possible copying mechanism for the genetic material.'

This celebrated understatement published in *Nature* on 25 April 1953 was Francis Crick's and James Watson's way of heralding the significance of their discovery of the double helix, the self-copying spirals of the DNA molecule that carry the genetic message from old cells to new. Another statement, written in a private letter on 7 March 1953, has achieved a fame of its own: 'Our dark lady is leaving us next week.'

For Francis Crick of the Cavendish Laboratory at Cambridge, the 'dark lady' needed no further identification. For nearly two years, his friend Maurice Wilkins of the Biophysics Unit at King's College London had been moaning about his obstructive female colleague, Rosalind Franklin. Now that she was abandoning King's for Birkbeck, another University of London college, Wilkins was confident that he, Crick, and Watson, a young American working with Crick, together would solve the structure of DNA. But it was too late. By the time that Wilkins's letter reached Cambridge, the pair whose names will be forever linked were looking at their completed model whose simplicity proclaimed that they had discovered the secret of life.

But could Watson and Crick have done it without the 'dark lady': Rosalind Franklin, the thirty-two-year-old physical chemist whose departure from King's Wilkins so eagerly awaited? Her research data, which had reached them by a circuitous route and without her consent, had been crucial to their discovery. Watson's glimpse of one of her X-ray photographs of DNA gave him and Crick the final boost to the summit. From the evidence of her notebooks, it is clear that she would have got there by herself before long.

The triumph was theirs, not hers. Rosalind Franklin remained virtually unknown outside her immediate circles until 1968 when Watson published *The Double Helix,* his brilliant, tactless and exciting personal account of the discovery. In it, she is the terrible 'Rosy', the bad-tempered bluestocking who hoarded her data and might have been pretty if she had taken off her glasses and done something interesting with her hair.

She looked quite different to the eminent physics professor J.D. Bernal, who brought her to Birkbeck in the spring and oversaw her five happy and productive years there. He described her in *Nature*: 'As a scientist, Miss Franklin was distinguished by extreme clarity and perfection in everything she undertook. Her photographs are among the most beautiful X-ray photographs of any substance ever taken.'

But Bernal's words were elegiac. Rosalind Franklin's life was cut short by ovarian cancer in 1958 when she was thirty-seven – four years before Watson, Crick and Wilkins won the Nobel prize for their DNA discovery and a decade before she was caricatured in a book to which, alone of the principals portrayed, she was unable to answer back.

Since Watson's book, Rosalind Franklin has become a feminist icon, the Sylvia Plath of molecular biology, the woman whose gifts were sacrificed to the greater glory of the male. Yet this mythologising, intended to be reparative, has done her no favours. There was far more to her complex, fruitful, vigorous life than twenty-seven unhappy months at King's College London. She achieved an international reputation in three different fields of scientific research while at the same time nourishing a passion for travel, a gift for friendship, a love of clothes and good food and a strong political conscience. She never flagged in her duties to the distinguished Anglo-Jewish family of which she was a loyal, if combative, member.

Determined from the age of twelve to become a scientist, Rosalind Franklin knew where she came from, under what constraints she laboured and where she wanted to go. From childhood, she strove to reconcile her privileges with her goals. She did not find life easy – as a woman, as a Jew, as a scientist. Many of those close to her did not find her easy either. The measure of her success lies in the strength of her friendships, the devotion of her colleagues, the vitality of her letters and a legacy of discovery that would do credit to a scientific career twice its length.

'You look at science (or at least talk of it) as some sort of demoralising invention of man, something apart from real life, and which must be cautiously guarded and kept separate from everyday existence. But science and everyday life cannot and should not be separated. Science, for me, gives a partial explanation of life. In so far as it goes, it is based on fact, experience and experiment . . .

I agree that faith is essential to success in life . . . In my view, all that is necessary for faith is the belief that by doing our best we shall come nearer to success and that success in our aims (the improvement of the lot of mankind, present and future) is worth attaining.'

Rosalind Franklin as a Cambridge undergraduate arguing against her father's faith in life after death.

Part One

ONE

Once in Royal David's City

THE FAMILY into which Rosalind Elsie Franklin was born on 25 July 1920, stood high in Anglo-Jewry. Not at the very top: the highest rank was occupied by the oldest Jewish families in England, the Sephardi Jews of Spanish and Portuguese descent who arrived at the time of Cromwell. Nor were the Franklins among the wealthiest of the Ashkenazis from northern Europe, such as the Rothschilds and Goldsmids, who came to England in the eighteenth century seeking opportunity for trade. Yet they were well within the elite network known as 'The Cousinhood', so common was intermarriage.

The first of their English line arrived as Fraenkel from Breslau in Silesia in 1763 and anglicised the name to Franklin, as was sensible. The English were uncomfortable with foreign names, and Jewishness was no advantage at a time there were only 8,000 Jews in England. Benjamin Wolf Franklin lived in the City of London, on Cock Court, Jewry Street. A rabbi and teacher, he married Sarah, the daughter of Lazarus Joseph, originally Lazarus Israel, who emigrated to England from Hamburg around 1760. Benjamin and Sarah had six children before dying in an epidemic in 1785.Their gravestones still stand in the burial ground in Globe Road Cemetery, Mile End, East London.

The two surviving Franklin sons, Abraham and Lewis, went to Portsmouth for apprenticeships in watchmaking and shop-keeping and became successful businessmen. In 1815 or 1816 the brothers shifted to Liverpool and Manchester where they entered firms engaged in money-changing, banking and trade with the West Indies. In time, the Samuels of Liverpool (another line of Silesian exiles) were braided into the Franklin fabric. Over the generations the Abrahams became Alfreds and Arthurs, and the ancestral surname Israel became Ellis. In 1852 the grandson of the original immigrant, Ellis A. Franklin, joined Louis Samuel from Liverpool in the bullion-broking firm of Samuel Montagu and Co., and the alliance was cemented when Ellis married Samuel Montagu's sister.

From 1868 the Franklin family's financial base lay in the City of London, in A. Keyser and Co., a private merchant bank spun off from Samuel Montagu and Co. Keyser's, a source of employment for Franklin sons for the next century, became independent in 1908 and specialised in placing American rail bonds in the City. Among the City's so-called 'Jewish banks', Keyser's was the only one to observe all the Jewish holidays.

From 1902 Franklins were publishers as well as bankers. Keyser's bought the house of George Routledge from the receivers in 1902 and in 1911 took over another ailing publisher, Kegan Paul. This acquisition created a refuge for Franklin males disinclined to banking.

In 1862, in line with the exodus from the City of London where the Jews had clustered, Ellis Franklin shifted to west London and the wealthy Jewish enclave in Bayswater. There he was one of the founders of the New West End Synagogue on St Petersburgh Place. His seven children, all of whom married within the faith, made the family known for prodigious philanthropic zeal – a leading example of the Jewish tradition of repaying the privilege of wealth through service to those less fortunate. In succeeding generations there was scarcely a Jewish organisation, hospital or old people's home without a Franklin on the board and many secular charities benefited from their dedication as well.

On her mother's side Rosalind's antecedents were intellectual and professional. The Waleys had been in England even longer than the Franklins, having arrived in Portsmouth in 1740 as Levis. Rosalind's maternal great-grandfather, Jacob Waley, took first place in mathematics and classics at University College London, and later became professor of mathematics at the University of London while also practising at the bar. He too was active in Jewish good works, as a founder of the United Synagogue (an association of nominally Orthodox synagogues which observed the German or Polish ritual), and first president of the Anglo-Jewish Association. In a prime example of 'The Cousinhood' in action, he married Matilda Salomons, a niece of both Sir Moses Montefiore and Sir David Salomons, the first Jewish Lord Mayor of London.

Anglo-Jewry was a happy breed: secure, able, influential, socially conscious, cosmopolitan. Its members dressed for dinner, were presented at Court, had their portraits painted by Singer Sargent. Many kept Christmas and Passover, ate kosher and played cricket. The price of belonging was intermarriage. But there was no need for exogamy – the dreaded 'marrying out'. As prolific as other families of the late-Victorian and Edwardian eras, the clans produced enough offspring to stock a generous marriage pool, and to speed the path to it through a well-organised social round of dances, picnics, theatre parties and country weekends. Rosalind's parents, Ellis Franklin and Muriel Waley, met at an engagement party at the family house in Porchester Terrace, Bayswater, and began their courtship by escaping for a long walk across Hyde Park.

Which was the stronger loyalty – to country or to faith? There was nothing to choose. As Rosalind's father said when reorganising the New West End Synagogue after the Second World War, 'The whole idea is that Judaism is a religion not a race . . . the English Jews are as much English as other English.'

The discovery of the secret of the gene involves a genealogy as long as any in the history of the planet. Elders of the Franklin clan claimed direct descent from King David, founder of Jerusalem, reigning king of Israel *c.*1000 to *c.*962 BC, around whom the messianic expectations of the people of Israel clustered. The lineage was spelled out by Rosalind's grandfather, Arthur E. Franklin, in a thick blue book beautifully produced by Routledge, the family firm.

'It may appear strange,' he conceded in his introduction, 'that anyone living in the fourteenth [actually fifteenth] century could trace his descent from King David.' He then proceeded to lay out the evidence. Name by name, date by date, he followed the Franklin blood line back through the Bohemian Jewish communities of the Holy Roman Empire to the great Rabbi Löwe of Prague, who died in Prague in 1609, a scholar, writer (allegedly the creator of the Golem, a supernatural monster), scientist and friend of Tycho Brahe, the astronomer. From there he followed the trail farther back to another ancestor, the Imperial Rabbi of Prague who died in 1439.

Arthur Franklin acknowledged that there was a large gap between 1038 and 1439 where the record was missing, probably destroyed in a fire in Prague in 1689. However, from other books on the history of the Jews in Bohemia and with help from the synagogue library at Breslau, he had reconstituted the missing links with the Exilarchs of Babylon, rulers of the Jews expelled from Jerusalem in 587 BC after the fall of the First Temple. Thus the Franklins, as he saw them, were descendants of the Exilarchs. The office of Exilarch was always held by a descendant of the House of David. Therefore, in the second edition of *The Franklin Family and Collaterals*, published in 1935, he confidently placed Jehoiachin, the first Exilarch of Babylon, on page 67, and Rosalind Elsie Franklin, second child and first daughter of Ellis and Muriel (Waley) Franklin, on page 85.

The intellectual component of this genetic inheritance was clear: the Franklins came from a line of scholars and leaders. That said, what mattered just as much to Arthur Franklin was that the 3,500 entries recorded in his family history represented 'a fair proportion of the English Jews whose families were settled here [England] before the Napoleonic wars'. His own proud boast was that three out of his four grandparents were of English birth.

Old Jewry indeed. In class-stratified Edwardian England, eighteenth-century origin placed the Franklins in the upper middle class, high above the new wave of Jews crowding into London's East End in flight from the pogroms in Russia and Eastern Europe. Thus the Franklins were archetypal – to use a term they did not use of themselves – assimilated Jews.

By the time Rosalind was born in 1920, the Franklin family and collaterals stood high in British public life. Three weeks before her birth, the great-uncle she would call 'Uncle Herbert' was installed in Jerusalem as the first High Commissioner of Palestine. So well established was Herbert Samuel that he received the mandated territory from its military commander in the form of a light-hearted receipt: 'Handed over to Sir Herbert Samuel, one Palestine, complete.'

For Herbert Samuel of Liverpool, who in 1897 had married his cousin, Beatrice Franklin, the Palestine commission was one more step in a distinguished political career. In 1909 when appointed chancellor of the Duchy of Lancaster in H.H. Asquith's Liberal government, he became the first practising Jew to sit on the British Cabinet. (Although Jewish, Benjamin Disraeli, the Victorian prime minister, was a baptised Anglican.) Samuel rose to become Postmaster General, then in January 1916, Home Secretary, a post he held for a significant year which included the Easter Rising in Dublin and the summary execution of the rebellion leaders. Samuel had no natural sympathy for national aspirations of colonised people when these clashed with British interests or loyalty to the Crown. He resigned in December 1916 upon the formation of David Lloyd George's wartime coalition government.

Before leaving office Samuel wrote the memorandum that resulted in the Balfour Declaration in 1917. In 'The Future of

Palestine', he outlined a plan whereby after the war and the presumed collapse of the Ottoman Empire, Britain would assume a protectorate over Palestine and encourage Jewish immigration until a majority was reached and Palestine could be granted self-government. (Jews at the time comprised only one-sixth of the population.) Explaining his reasoning, Samuel declared: that the link between Palestine and the Jews was as old as history, that Jews all around the world would be grateful (Britain was hoping to bring the United States into the war) and (with a nod to the eugenics theory popular at the time) that 'The Jewish brain was a physiological product not to be despised.'

However, Samuel accepted assignment as Palestine's first High Commissioner with a heavy heart. Like much of Anglo-Jewry, he was uneasy about the so-called national homeland, for Britain was their home. He had taken pains to make it clear that nothing should be done in Palestine 'which may prejudice the rights and political status enjoyed by Jews in any other country'.

English Jews fought the Jewish stereotype with incomplete success. Disraeli never threw off the shadow, despite baptism, a Christian wife and the efforts of a father who had removed the foreign-looking apostrophe from the family name, d'Israeli. 'Somehow,' a parliamentary sketchwriter observed in 1871, 'Englishmen have never yet been able to give their confidence to anyone who bears the unmistakeable traces of Jewish origins . . . had he been of British descent, like his lifelong rival Mr Gladstone, everything would have been forgiven.'

Disraeli's cleverness and the charm he held for Queen Victoria were ascribed to his Jewishness. So too was Herbert Samuel's ingenuity in designing the Balfour Declaration. Asquith described Samuel's plan to his friend Venetia Stanley as 'a curious illustration of Dizzy's favourite maxim that "race is everything"'.

'The Jew' was part of the cultural landscape in the country which in 1290, after prolonged harassment, had expelled them as well-poisoners, liars, usurers and baby-eaters. Dickens, in *A*

Child's History of England, wrote of their plight at the time of Edward I:

> in this reign they were most unmercifully pillaged. They were hanged in great numbers . . . heavily taxed; . . . disgracefully badgered . . . thrown into beastly prisons until they purchased their release . . . Finally, every kind of property belonging to them was seized by the King . . . Many years elapsed before the hope of gain induced any of their race to return to England, where they had been treated so heartlessly, and had suffered so much.

Cromwell allowed Jews back in 1656 after the Civil War. With the dawn of capitalism and the spread of trade the prosperous among them were welcome – up to a point. When the 'Jew Bill' of 1753 was introduced to make it easier for them to become naturalised British subjects, there were anti-Jewish riots and the bill was withdrawn.

They survived by becoming more English than the English, doing nothing – no loud voices, extravagant hand gestures, bright colours or conspicuous consumption – to evoke the 'bad Jew' qualities of Shylock, in *The Merchant of Venice,* written when Jews had been absent from England for 300 years.

Their numbers were still small (20,000–30,000) at the beginning of the twentieth century when the new wave of immigration threatened Anglo-Jews' hard-won acceptance by the host society. Charitable institutions were swiftly established to prevent those seen as 'not our kind of Jew' from seeking public welfare or otherwise stirring up the anti-semitism latent in English life.

Assimilation, however heartfelt, never looked complete. The hint of something eastern, alien and untrustworthy was carried in the strange language, with its curving oriental script, used for what the composer Richard Wagner, an ardent anti-semite, called 'the hidden discourse of the Jews'. The Jewish insistence on marrying within the tribe contradicted their claim to have adapted to the dominant culture. Intermarriage could also be read as inbreeding, with even darker connotations of insanity.

It was hardly possible to get through English schooling in the twentieth century without knowing of Shakespeare's Shylock, with his 'Jewish heart', moaning over his lost ducats, or of Ivanhoe choosing the fair Saxon Rowena over the raven-tressed Rebecca the Jewess.

If behaving well was the best defence, to the unsympathetic eye it looked like deception. Success and wealth were mixed blessings. Too much of either invited charges of avarice, materialism and profiteering. The tireless philanthropy of affluent English Jewry could be interpreted as looking after their own kind, or at very least, a defensive protection of position. The easy internationalism of the Jews was seen as a conspiracy, a desire for profit that had perhaps fomented the First World War. A character in *The Thirty-Nine Steps,* written in 1915 by John Buchan, who was director of information for the Lloyd George wartime government, maintains 'The Jew is everywhere . . . with an eye like a rattlesnake. He is the man who is ruling the world just now . . .'

Examples of the sophisticated anti-semitism that flourished in between-the-wars English letters are all too easy to find. Rebecca West, writing to a friend, described a young film executive as 'a very charming, very attractive young Jew of about twenty-eight – not a Kike, he belongs to a very old-established Jewish family'. The writer and diplomat Harold Nicolson argued against letting Jews into the Foreign Office: 'Jews are far more interested in international life than are Englishmen, and if we opened the service it might be flooded by clever Jews.'

In this harsh light an English surname like Franklin could look like a trick. In 1922 the Catholic writer, humorist and anti-semite Hilaire Belloc (*not,* however, as is widely believed, the author of 'How odd/of God/To choose/ the Jews') suggested that one of the causes of anti-semitism was the Jewish 'assumption of false names and the pretence of non-Jewish origin in individuals'. T.S. Eliot was also on the look-out. In his 1995 study of Eliot's anti-semitism, the Anglo-Jewish lawyer and writer Anthony Julius says, 'Rooting out the Jew behind the Anglicised name was a party game for the Jew-conscious; thus Eliot wrote to [Ezra] Pound, "Burnham is a Jewmerchant, named Lawson (sc. Levisohn?)."'

Julius argues that British anti-semitism has survived into the twenty-first century: 'Jews succeed against the grain. There is a certain resistance to them that is rarely expressed, and never legislated. Anti-semitism in England breeds Jewish paranoia. You don't see it coming; and when it's gone you're still not sure quite what it was. To understand what is going on in England, you need a very nuanced sense of the anti-semitic.'

One member of the Franklin family with a highly nuanced sense of the anti-semitic was Rosalind's aunt, Helen Bentwich, Ellis Franklin's sister known to the family as 'Mamie'. Already in Palestine when 'Uncle Herbert' arrived in 1920, Mamie was the wife of Norman Bentwich, the British-appointed Attorney General, one of the most powerful posts in the British Mandated Territory. She found her position as diplomatic hostess awkward as so many visitors from England were 'so very outspokenly anti-Jewish . . .'

> The Ashbees dined here, & he said that the Press in England is entirely controlled by the Jews, & only express Jewish interests! They refused to listen to my accounts of the *Morning Post* anti-semitism, & Northcliffe & Cadbury & Beaverbrooke & co.

Another visitor was the Conservative politician Stanley Baldwin's elder son: 'a young man with long yellow hair', who 'frankly said he hated all Jews – I told him off for being a foolish youth with popular prejudices, & we thrashed it out till very late hours'.

Mamie's determination to avoid ostentation came into conflict with her duties as hostess at the many formal engagements at

Government House. To her mother she wrote, 'I'm sure we don't splash – of course, the gold dress & fur coat look rich, & we make a habit of never talking about money.'

A classic example of the Anglo-Jewish predicament was furnished by Rosalind's forebear, Sir David Salomons, London's first Jewish Lord Mayor. Elected to Parliament from Greenwich in 1852, Salomons was set to become the first Jew to sit in the House of Commons. At the swearing-in ceremony, he declined to utter the Christian oath on the English Bible. Gracefully he alluded to the 'unusual course' he had taken:

> I trust the House will make some allowance for the novelty of my position . . . I thought I should not be doing justice to my position as an Englishman and a gentleman did I not adopt the course which I thought right and proper of maintaining my right to appear on the floor . . . and stating before the House and the country what I believe to be my rights and privileges.

The novelty of her position, as an English gentlewoman and a Jew, shaped Rosalind Franklin's life. Educated and trained in England, she moved to Paris in 1947 for a post in a French government crystallographic laboratory. She worked and lived there happily until 1950 when, at the age of thirty, she decided to accept an appointment in the new field of biophysics at King's College London. Reluctant to return, she hurt her patriotic parents deeply by writing 'I like Europe and the Europeans so much better than England and the English.' When they protested, she apologised, then said it all over again:

> Dear Mother, I'm sorry if you find my last letter distressing. It didn't really reflect any change of attitude, nor any bitterness either. I have always preferred 'foreigners' to the English.

Her proud but heavy heritage had left her uneasy in her native land.

TWO

'Alarmingly Clever'

IN SPITE OF and perhaps because of being the youngest of six children, Rosalind's father, Ellis Franklin, was a natural patriarch. Large, intelligent, amusing, successful, overbearing, he knew what was best for those around him and told them so. He could be quite angry if resisted. His service in France in the First World War as a captain in the King's Own Yorkshire Light Infantry Regiment was the defining experience of his life. Before war broke out, he had intended to study science at New College, Oxford. When he was released from the service in 1918, however, having married, he accepted the family's judgement that he must give up thoughts of Oxford and instead join Keyser's bank where his father was senior partner.

Ellis worked as hard at recreation as at business and public service. Married in 1917, he took his young wife on strenuous backpacking holidays in France and also in England, where they tramped over the Berkshire and Wiltshire Downs. Even in London they walked. Ellis would lead Muriel, a small figure struggling to keep up, on marathon treks from Notting Hill to Richmond and Wimbledon.

His children appeared in swift succession, the first, David, born in 1919. The family lived in a large double-fronted four-storey house at 5 Pembridge Place off Westbourne Grove in London W2 on the western edge of Bayswater, now fashionable as Notting Hill. To wealthier and grander Jewish families who lived in the great seven-storey stuccoed mansions of South Kensington, it was perhaps the wrong side of the park. Otherwise, the leafy squares and stuccoed terraces north of Kensington Gardens and Hyde Park sheltered a comfortable and affluent community where, in the words of a distant cousin, L.H.L. Cohen, 'we lived like Jewish Forsytes'. Nannies of the neighbourhood wheeled their prams to Kensington Gardens, watched their charges romp around the statue of Peter Pan and the Round Pond, and escorted them to each other's parties. The focus of the community was the New West End Synagogue on St Petersburgh Place – or, for some, the West London (Reform) Synagogue near Marble Arch; for observant Jews, it was important to walk (or at least not to be seen to drive or be driven) to synagogue on the Sabbath. Families also liked their children to live close by, so that they might dine together on Friday evenings when the Sabbath candles were lit, and on Passover. Ellis and Muriel Franklin were within easy reach of his father's house on Porchester Terrace and that of her parents, in Norfolk Square. The London home of Sir Herbert Samuel (a witness at their wedding) was also on Porchester Terrace.

For family holidays Ellis chose the Scilly Isles, Cornwall or South Wales, where, trousers rolled up, he would splash about with his brood, directing them where to look and what to explore. In 1926, when Rosalind was six, they were joined on the Cornish coast by Mamie Bentwich, his favourite sister. Mamie, who was childless, described the scene to 'Dearest Norman', her husband the Attorney General, back in Jerusalem:

> We are enjoying life here very much . . . walking, surf-bathing & lazing on the heather or sands – & eating a terrible amount.
>
> Friday Ellis & I started for a long walk – but it rained heavily so we spent all day indoors, except for an after-tea walk to a 7th century 'theatre' near here on the cliffs . . . Today, bathing & digging & playing with the children as

> the nurses were out. They are great fun . . . especially Colin . . . David is a charming little boy. Rosalind is alarmingly clever – she spends all her time doing arithmetic for pleasure, & invariably gets her sums right.

'Alarmingly clever': the phrase, affectionately meant, has a contemporary resonance inaudible at the time. 'Alarmingly' is not synonymous with 'amazingly'. In 1926 British women had enjoyed the vote for only eight years, and then only if they were thirty or over. Mamie, born in 1892, knew all too well that superior intelligence in a female was an embarrassment. She held a diploma from Bedford College, University of London, yet had been brought up in the understanding that neither she nor her older sister Alice would have careers. Their model was their mother, Caroline Jacob Franklin, also holder of a Bedford diploma, who used her intellectual gifts as chair of the Buckinghamshire Education Committee, as a member of the Anglo-Jewish Association, and as a supporter of the Evelina de Rothschild School in Jerusalem.

Alarm at Rosalind's ability was therefore appropriate. It was awkward that, of Ellis Franklin's four children, three of them boys, the girl should be the brightest and most determined. Ellis himself would not allow women employees in his office; for years Keyser's bank had only male secretaries and telephonists.

That Rosalind at six should be so adept at arithmetic was owed in part to a good west London private day school. Like her older brother David, she was a pupil at Norland Place on Holland Park Avenue. Many mornings Ellis walked them the mile or so from Pembridge Place to the school which was – and still is – situated in a row of small four-storey houses. Norland Place was unusual for its time in offering coeducation until the age of eleven; after that it taught girls only. For five and a half guineas per term, it gave a good grounding in history, literature and arithmetic. It was a progressive school in that it taught woodworking, called 'Sloyd', to girls as well as boys. 'Sloyd' was a Swedish method of training children in the practical beauties of wood. Norland also took a unisex approach to sports; both boys and girls were encouraged to play cricket and hockey – sports Rosalind much enjoyed.

A group photograph was taken on the playing fields in 1926 on the occasion of Norland Place's golden jubilee. It shows a row of women teachers, many in cloche hats, beaming broadly, while in front, seated on the grass with the smallest pupils, a serious little girl, wearing spectacles, is in no mood for putting on a silly grin and looks straight at the camera with a direct unsmiling gaze.

Rosalind had a loving mother. In a sense, she had two. Muriel Franklin was a gentle, well-educated, intelligent woman who played (and, some of the wider family felt, over-played) the role of subservient Jewish wife according to the tradition in which woman was synonymous with wife and mother as defined in Proverbs 31:28: 'Her children stand up and honour her, and her husband sings her praises.' In that part, Muriel was exemplary. Highly intelligent and able, in later years an accomplished writer, she was denied a university education through her mother's disapproval. As a married woman she threw herself into charity work under her husband's direction. She taught her children good manners and the supreme importance of family loyalty. She insisted on respect for grandparents, which meant writing a thank-you letter for every present as soon as it was received; also inquiring about the old people's health and reporting progress at school. One of Rosalind's earliest letters reads, 'Dear Grandma – I got two stars at school.'

In rearing her large brood and assisting her husband with his myriad charities, Muriel was able to delegate day-to-day authority over the children to a loyal nanny, who came for the first baby and stayed until the last, and long after. For Rosalind, in a life that was in many ways a battle, Ada Griffiths was a source of warmth and refuge. A robust woman from Shropshire, 'Nannie' as the children called her (and usually spelled it), like her two sisters, accepted looking after other people's children as the only career open to her. She had an additional obstacle to overcome: two club feet from which two toes had been amputated. The

Franklin children used to watch fascinated while she wrapped her maimed feet in bandages to fit inside her sturdy surgical boots. Muriel and Nannie got on well in their mutually advantageous relationship. To Nannie, a Franklin cousin observed dryly, 'the five Ellis children were all swans'.

Nannie found her purpose in life with the generous employers with whom she cast her lot. With her sweet smile, her practical ability to deal with cuts, bruises, tangled hair and sibling squabbles, she was the centre of the Franklin children's lives. As one of them later said, 'Mother was for extras.'

Nannie was so much part of the family that her birthday, like one of their own, was an occasion to be celebrated and planned for. As Rosalind wrote her grandmother, 'It is Nannie's birthday to-morrow and we are all getting excited, Roland has bought her something which was a secret and painted a gourd which she knows about and she does not know he has got anything else.' When Nannie went home on holiday to visit her family in Church Stretton, Rosalind's constant plea was 'When is Nannie coming back?'

Nannie taught Rosalind to knit, to be a perfectionist about her handiwork and to succeed in her intention. Yet even she could favour the boys. Once Rosalind complained that Colin had hit her with a cricket bat and pleaded to Nannie, 'Do something!' Nannie's cheery response was 'Well dear, you shouldn't have been teasing him.'

That Rosalind's mother should mention this teasing streak in a loving posthumous tribute to her daughter suggests that the trait was quite strong: 'One recalls a smug, naughty-sounding little voice in the darkened night nursery where, with her still small though older brother – who adored their nanny – she was supposed to be asleep. "Nannie is a dustbin, Nannie is a boiled egg. I'll take Billy [their canary] and cut him into bits and put him in a sandwich. You wouldn't like that, would you, David?" She was not quite two years old.'

All her life, Muriel recalled, 'Rosalind enjoyed this trick of teasing – but with a burbling, mischievous delight, that, for anyone with a sense of fun, took the sting out of it. Only occasionally when she was vexed for some reason, or in a contrary mood, would the teasing become provocative – a tantalising naughtiness that could be quite maddening.'

Indeed, Rosalind was a stormy child, easily roused to tears and anger. Her family never persuaded her she had equal treatment with her brothers, particularly with David, the eldest, so close in age that they had shared a double pram. If her family was a little afraid of her, so was she of them. One of her early drawings is captioned in shaky big print: 'A child being Lifted Up by the Hair'. The child is a girl with brown hair and a yellow patterned dress, yellow socks and black shoes. The woman shown in profile has a spotted green and yellow dress, bright lipstick, a green eye and a red tongue stuck way out. She looks angry. Who was it? Not Nannie. Not Muriel Franklin, who struck those who did not know her well as being timid as a mouse.

The Franklins were frugal rich. The joint at Saturday lunch was sliced thinly and the leftovers kept for re-serving. Ellis made his wife provide meticulous accounts for the housekeeping, down to the last penny as if home were part of the bank. The children were taught to care for small sums and to pay scrupulous attention to the cost of anything they bought.

Theirs was hardly boundless wealth. They were well-off rather than very rich. For many years Keyser's total profits were about £200,000 a year. Grandfather Franklin did live in the grand style: his townhouse, and Chartridge Lodge, his country estate in Buckinghamshire, were modernised by the same architect who added the facade to Buckingham Palace. Unlike his father, however, Ellis Franklin did not want a second home nor a chauffeur. He went by Underground from Notting Hill Gate to Bank. The one indulgence was foreign travel.

It was a habit acquired from his parents. The Arthur Franklins travelled widely, through the Middle East, Scandinavia, Serbia,

France and Italy, all without relaxing their orthodox practices during their travels. They went without meat rather than risk the chance of eating a forbidden food.

Ellis on his own, after finishing his public school – Clifton College in Bristol – spent a year in Breslau (now Wroclaw in Poland, a city on the Oder between Prague and Warsaw) where there were still family connections. There he studied science and became fluent in German, as a preparation for the scientific career he was not to have. When he returned from Breslau ill, he was sent on a round-the-world sea voyage with his older sister Alice as chaperone. Travel in itself carried responsibilities: long detailed letters home showing what had been gained from it. Ellis's letters from Australia and New Zealand were kept in the family archive, and probably read aloud. He returned in July 1914, just before war broke out. He enlisted immediately.

For Rosalind, the exhilaration of foreign travel began early. Her Franklin grandfather had suffered from tuberculosis in 1925 and, with his wife, spent winters at Menton on the French–Italian border. During the second of these Ellis and Muriel took their two oldest children to visit them. Since childhood Ellis had loved riddles and crosswords, and as a father worked to sharpen the minds of his own brood. He turned each stage of the trip – P&O liner to Marseilles via Gibraltar, then train – into an improving game. On the ship he led the children every morning in skipping energetically on the deck, taking care to skip as long and hard as they did. As for the train journey, as Rosalind explained in her prompt thank-you, from the Hôtel Splendide, Marseilles (in childish print on paper with hand-drawn lines):

> Dear Grandma,
> Thank you very much for asking us. In the train, we had a game. Daddy put a lot of things – Pen Pencell Ring bag Coin Pen Knife Note book Cigarette case and Papper on a shelf took it off and we had to look and when he put it on again we tried to see who could remember the most things.
> With love from Rosalind.

In a later letter, Rosalind reported, 'They are having a bazaar at the College for which we have started making things even Roland.' (Roland, the youngest, was four at the time.) 'The College', like 'the Bank', was a cornerstone of the Franklin children's existence. The Working Men's College, a Victorian redbrick institution stretched along Crowndale Road between Regent's Park and King's Cross Station, was Ellis Franklin's favourite charity. Into it he poured his thwarted scientific talents and gave long unpaid hours two or three evenings a week on his way home from the City. He taught electricity, magnetism, and the history of the Great War, and became in succession director of the Science Group, treasurer, and vice principal. As one of its goals was to heal the divisions between the educated and the uneducated, Ellis stayed on for conversation and chess after classes. He rarely reached home before 11.30 p.m. on teaching nights.

The College was founded in 1854, by Frederick Denison Maurice, a Christian Socialist, to bring together 'men of the working classes and men from the Universities, in the common object of teaching and learning'. John Ruskin was one of its earliest teachers. After 1917, this ideal encompassed the hope, never spoken aloud, of staving off Bolshevist revolution through befriending and educating the labouring classes before they got hostile ideas.

The principal in Ellis Franklin's time was Major-General Sir Frederick Maurice, grandson of the founder. For his part, Ellis used his financial acumen to raise the College's endowment to an affluence beyond the founder's dreams. With the help of City friends, he bought and had developed a sports ground for the College in Edgware on the northern outskirts of London. For the children the College's annual sports day was second only to Christmas and Jewish religious holidays as a command performance for which they were expected to do their best – not least by making sandwiches. However, Ellis agreed with the governing body that women would not fit into the life of the College. Accordingly, no women – not even Muriel Franklin, who had volunteered to dust the books – were allowed in the College Library.

There was a role, however, for the pretty three-year-old daughter of the vice principal. In 1923 Rosalind was chosen to present a bouquet of carnations to the Duchess of York at the opening of the annual bazaar. The Duchess (later Queen Elizabeth and still later, the Queen Mother) complimented the College on its eighty voluntary teachers and one thousand new students united in their endeavour 'to give the working man of London a chance of entering into real college life, and to bring all classes together in pursuit of education'. For the occasion, Rosalind wore a beautiful lace-trimmed dress and had her shiny black hair well-brushed. Her intense interest in her clothes was already manifest when (as her mother recalled) at the same age she tore a frock because it had no 'ribbings' in it.

Rosalind's childhood, therefore, was spent in a settled world of nursery, school, park, pets, pantomimes, birthday parties, holidays, sports days and country weekends at grandfather's. In 1929 it all changed. A new baby was on the way.

> Dear Mummy and Daddy,
> Please tell Roland we had jam pudding the other day. It is lucky Miss Falckes Lesson [the Franklin children's Hebrew teacher] did not come as they did not go to church. The first few nights we were five in the dormatory but I have been moved and now we are only four. The oldest in the room is twelve . . . We go to bed at quarter to seven. There are 10 in my form, I am the youngest, it is the second form.

The bracing air of the Channel coast was the family's reason for sending Rosalind, at the age of nine, to boarding school. But her dispatch from the family nest and the comforting care of Nannie coincided with the arrival of the new baby. The Franklins' fifth child and second daughter, Jenifer, was born on 5 November 1929, a welcome break in the wall of brothers which surrounded Rosalind. The school Ellis and Muriel Franklin chose for their now-elder daughter was at Bexhill on the Sussex coast, an area crammed with hotels, boarding schools and golf courses, in the belief, not groundless, that the seaside offered a healthier environment than the smoke and coal fires of London. David, the eldest boy, was at school nearby and the parents persuaded themselves that Rosalind, who had had several childhood illnesses followed by periods of forced rest, was frail and needed the change.

Visiting, her mother was satisfied that the school was curing Rosalind's weakness: she was brown, healthy and worried only about the fatness of her legs. As Muriel respectfully reported to her father-in-law: 'She is obviously very happy & leading a healthy sensible life. Her work too seems to be on sound lines & very interesting. They do sewing in their spare hours & she is proudly making Jenifer a frock – & very nicely too.' Muriel was satisfied to see the girls at the school 'playing & lying about the garden – it looked very peaceful & happy'.

Poignant letters show that Rosalind would have liked the pleasure of watching her tiny sister grow up and that she longed for home ('What is the new kitten like?'), but she accepted exile with the brisk alertness she had brought to life with three brothers. Driven in on herself, she used her intelligence and self-control not to show emotion, but instead poured her feelings into her work and handiwork, in which she showed an extraordinary facility to align mind and hand.

Lindores School for Young Ladies, founded in 1912, was no grim pile. A half-timbered Victorian mansion with dormer windows and a long sloping lawn, it was the former home of a diplomat and had a rambling residential air. Yet the pleasures it offered were spartan: a half-mile brisk walk to the flagstaff near the colonnades on the seafront and, in the evening, lantern-slide lectures: 'Travelling Across Canada', 'The British Empire', and one night: 'There was a lantern lecture chiefly on Palestine, and there was a large picture of Uncle Herbert, I recognised him before we were told who he was . . .'

As the school had no chapel, pupils attended St Peter's Church in the town. While 'they', as Rosalind's letters referred to the Christian girls, were at church, she had to do the Hebrew lessons sent to her through the post. If she did not complete these fully, her parents wanted to know why not. Her answers helped to fill the compulsory letters home: 'As they did not go to church again I did not have time to do much of the paper, I am sending it back. I missed out some qestions, because I forgot to bring a prayer book, please could you send one.'

But she liked the work, faster-paced than at Norland Place and taking her into new territories of geometry, geography and Elizabethan poetry. Her enthusiasms burst through: sports, handiwork, a fascination with foreign places, and above all, science:

> Dear Mummy and Daddy,
> There was a very interesting lecture last night, it was on the different way of a river beginning. Their were pictures the whole time, something like a cinema . . . First he showed us some wonderful pictures, mostly taken from the air of the clouds and told us what weather, all the different kinds meant. Then he showed us some of the Norwegian mountain tops with snow on them and glaciers and frozen lakes, lots of which he took himself, then he showed us some of tall rocks in Belgium, about 1,000 feet up, and absolutely straight with narrow sort of valleys, about ten feet broad, which had been worn from a stream . . . which had been running for hundreds of years, and of the Victoria and Niagara falls . . .

A landmark of the week was the award of marks every Monday morning. She strove to be first in the form, and in time, achieved it. Often she was placed no higher than third – sometimes unfairly.

> Somebody must have made some mistake in the adding up and percentaging of last week's marks, for Mrs Blundle got it to 78% instead of 88%. I had 5 tens, 5 eights, a good many nines and 1 seven – I know one seven was the only

> mark I had below 8, and Mrs Blundle agreed with that. I was really top, but got put down as fourth, which I cannot possibly have been, as two of the people who beat me did not get one set of marks all through the week higher than me.

But, in the traditional terror of boarding school or prison, she could not risk complaining against the regime in case her letters were read. 'Please do not say anything to any mistress about the wrong marks, as I should not really have been keeping my own marks at all . . .'

Every term she counted the days until the end and conquered her homesickness by making presents such as a bonnet for the baby and a knitted scarf for Nannie (made to her own meticulous colour-coded design which she sent to her parents).

By the end of the summer term, 1931, she was gaining top place or very near it almost every week. Her handwriting had matured, she had done a great deal of reading and she was skilled in hockey, swimming and drawing. She returned home, well-taught and self-sufficient. But she dismissed as nonsense her parents' belief that she had been 'delicate' and remained later in life resentful at having been sent away.

THREE

Once a Paulina

> 'At St Paul's . . . every girl is being prepared for a career. The High Mistress considers that no woman has a right to exist who does not live a useful life.'

MISS ETHEL STRUDWICK, the second High Mistress of St Paul's Girls' School, founded in 1904, inherited this philosophy from her predecessor and understood very well that it did not exclude motherhood as a useful occupation. However, she firmly expected the bright girls in her charge to look beyond marriage as their goal. Under her aegis, the school's magazine, *The Paulina,* ran a series on women in the professions, and a debate was held on the provocative motion, 'That the Entry of Women into Public Affairs and Industry is to be Deplored'. Miss Strudwick herself delivered the conclusion: women should not 'be relegated to the home'.

In January 1932, at the age of eleven, Rosalind entered the middle-fourth form of St Paul's, a west London day school known for its academic rigour. The school, on Brook Green in West Kensington, was a short bus ride from '5 PP' – as the family called the Franklin home at 5 Pembridge Place, Bayswater. Her Aunt Mamie had been one of the early pupils at the girls' school opened as a somewhat belated counterpart to St Paul's Boys' School, founded in 1506. Both schools, run by the Mercers' Company, a City of London guild, held no church allegiance and were accordingly a popular choice for London Jewish families with academically able children.

St Paul's was excellent for the competitive girl; it was not good at bringing out the violet half-hidden by a mossy stone. Rosalind was a natural Paulina. She would sweep home from school, swinging her satchel, wearing her round blue felt hat with turned-up brim and her blue school uniform from Daniel Neal. She threw herself into the team sports the school held dear – hockey, cricket and tennis – and in time she joined the Debating Society. On prize day – and she won a prize every year – she wore a white dress with white gloves. In the tradition of the English public school, her St Paul's friendships would last for life.

Although the girls' and boys' schools were just a few hundred yards apart, they were divided by more than the busy Hammersmith Road. There was no joint instruction nor shared activities, and little fraternisation. Even chatting together at the mutual bus stop was uncommon. Paulinas, like Paulines, took their work seriously. As one of Rosalind's contemporaries recalled, 'We weren't very interested in boys – we were interested in work and sport.' The rewards for diligence were evident. By Rosalind's time, Paulinas held prominent jobs in public life, in publishing, medicine, the law and the civil service.

A number of Rosalind's school friends were also her cousins: Livia Gollancz (later of the publishing company), Catherine Joseph and Ursula Franklin. But others were girls from different backgrounds and other parts of London, who, when invited to Rosalind's home, could be overwhelmed.

To Jean Kerslake (later Kerlogue), who lived in suburban Ealing, 'It was goggle-making': the heavy furniture, the maids answering the door and taking away the plates, the presence of Nannie and an under-nanny. Jean learned the rules of the household: Nannie was *not* a servant, she had a lot of authority over the children and lived at the top of the house. Jean was occasionally invited to stay the night, to join the Franklins for light opera in a box at the Albert Hall and to stay for Saturday lunch, when there was always roast beef and a lot of people around the table. She found Rosalind forthright, amusing, opinionated and adventurous, but was frightened of Ellis Franklin. None of the family suffered fools gladly: 'If you said anything silly, they would laugh.'

Another enduring friend was Anne Crawford (later Piper) from Putney. Both girls were always in the first division for mathematics, French, Latin and science – and both held scholarships and wore the scholar's badge on their tunics. Scholarships also covered tuition fees, for which Anne's family, professional but not well-to-do, were grateful. Rosalind's parents never took the money. Anne knew them to be 'a very very public-spirited family', and very well-off. Even so, she was startled when invited to spend a part of a summer holiday at St David's in Wales to find that the Franklins had taken a very large house and in addition to hiring local help had brought three of their maids with them.

Ellis Franklin did not want a country house, believing it was wrong to have two homes when many people did not have one. However, his parents' estate at Chartridge served the purpose. There were two tennis courts, a croquet lawn, a nursery lawn (with a swing and parallel bars), a kitchen garden, numbered bedrooms, a five-car garage with chauffeur's flat above it, a farm producing the household's own milk from Jersey cows, as well as chickens and turkeys and a resident *shochet* for kosher slaughtering in the approved manner.

Rosalind enjoyed weekends there with her Paulina cousin Ursula Franklin, the daughter of Ellis Franklin's older brother Cecil. Ursula liked Rosalind the best of all her 'Ellis' cousins, for her sense of humour and sparkle, and recalled the formality of Chartridge: 'It was always "Miss Rosalind" and "Miss Ursula" and you were not allowed to answer the door even if you were sitting near it. A maid had to come from the back of the house to open it.' Arthur Franklin, strictly orthodox, insisted on men wearing hats at table at Sabbath. He wore his yarmulke but his grandsons were more laconic: Colin sported a trilby while Roland wore his schoolcap.

Rosalind and Jean Kerslake became best friends in the spring term of her first year when each had returned from the Easter break having quarrelled with her previous favourite. Their joint antagonism towards Jean's discarded friend caught the school's attention and both sets of parents were summoned by the High Mistress to hear the complaint that their daughters were bullying Margaret Douglas. 'Rosalind and I achieved a certain notoriety,' Jean recalled, 'which strengthened our alliance.' This took them into joint projects, such as a small book on Japan, Rosalind doing the maps and drawings, and becoming patrol leaders in the Girl Guides, both winning among other proficiency badges, one in signalling – sending messages in Morse and semaphore.

Rosalind's parents were coming to recognise that their elder daughter could be difficult outside the family circle as well as within it. They sympathised with the efforts of her teachers to break through what Rosalind's mother called 'her reserve and apparent lack of response'. She referred to it as Rosalind's 'walking alone' stage. There were, as Ellis and Muriel well knew, strong rebels in the Franklin family. The most notorious was Ellis's brother Hugh, a pro-suffragist who in 1910 had accosted the then Home Secretary Winston Churchill on a train and had attempted to strike him with a dogwhip because of Churchill's opposition to women's suffrage. (Churchill was unharmed by the attack and continued on to the dining car.) For this episode, Hugh drew headlines in *The Times* as 'Mr Churchill's Assailant' and was imprisoned for several months. Then there was Ellis's older sister Alice. From a conventional start – Alice had been a debutante and presented at Court in 1909 – she turned into a left-wing socialist with cropped hair and pinstriped clothes, sharing a household with a woman partner in an arrangement about which the family made no comment. (Alice's intellectual energies were diverted into the Fawcett Society and Townswomen's Guild, for which service she was awarded the OBE.) As for Aunt Mamie, during the First World War, before she married Norman Bentwich and moved to Jerusalem and shocked the diplomatic corps by driving her own car at all hours, she had been dismissed from the Woolwich Arsenal for 'Bolshevist tendencies' – organising a trades union for women workers.

Was Rosalind going to be a younger version of these radicals? The teachers could not get through her apparent hostile indifference. As a strategic manoeuvre, Ellis and Muriel, acting either with imagination or on advice, bought Rosalind a Persian kitten. Named by Rosalind 'Wilhelmina' or 'Willy', the pet played its part in calming her down, waiting for her to come home from school and perching on the arm of her chair while she did her homework. What the nervous parents were probably witnessing was simply the beginning of adolescence. Rosalind's menstrual periods began when she was thirteen.

Trouble with teachers, anxiety over marks, delight in science, sport and sewing: the pattern from boarding school reasserted itself at St Paul's. In a sequence of revealing letters written to her parents when they were away, Rosalind poured out her schoolgirl woes, with confidence to write the way she spoke:

> Dear Mummy and Daddy,
> I was told in the drawing lesson this morning, that I was too literal-minded; We were doing designs for lino-cuts, they are lovely fun, and I did a thing something like this. [detailed diagram] She said that when there was a lot of it it would look too spotty, couldn't I put tails onto the rounds. I said I did not see how you could put tails on rounds so she said I was too literal-minded, and this was what she meant. This was the result. [another diagram] Would you have been able to guess what she meant? We spent the whole arithmetic lesson today with a lovely discussion about gravity and all that sort of stuff. I am doing my knitting on 4 needles, as Nannie said it would look untidy on two. I

> have come here at the right time as there is the party on Friday and we are probably going to the military tournament the following Friday . . .

Sadly, Rosalind never got to the Royal Tournament, held at Earls Court. Her next letter told the sorry tale:

> *I am very cross.* I was all dressed up in my hat and coat ready to go to the tournament last night, and the old pig came up to me and asked me what I was doing, she said I had told her I did not want to go, and she had given my ticket away so I could not go. I never said anything of the sort. Could I possibly go with anybody on Friday evening or Saturday, that will be the last opportunity as it stops on June 2nd. Everybody says this years is miles the best they have seen.

The following week the struggle continued: 'I only got B to B+ for that essay, which is very bad (I told you she was an old pig).'

Rosalind asked her parents' permission to take the examination for a senior scholarship. There was absolutely no danger of her getting it, she assured them. Even so, 'We are having gorgeous geography lessons, learning to weather forecast . . . We are going to keep records of the weather, clouds, signs, etc. for the next two weeks.'

Her pessimism was unwarranted. She won a Senior Foundation Scholarship, as she had won a Junior, and held it throughout her time at St Paul's. She also won the Latin prize in 1936. In English she disliked having to 'stodge' – her verb – through *School for Scandal* and *The Rivals,* and she did as little as possible of the music in which St Paul's prided itself, with the compcser Ralph Vaughan Williams replacing the late Gustav Holst as head of music in Rosalind's time at the school. She never pretended to have an ear or aptitude for music, although she did love theatre and painting.

* * *

At St Paul's Rosalind did not attend school prayers. What to do with Jewish pupils during morning assembly was a perennial problem for English schools. Even if they were independent of the Church of England, their day and year were structured around the ecclesiastical calendar and the statutory obligation to conduct a daily act of worship. In the 1830s when Rosalind's great-grandfather and his brother became the first conforming Jews to attend Manchester Grammar School, considerable ingenuity went into allowing them to be present at, but not participants in, school prayers. On the first day the Franklin boys were put in the front row and allowed to stand while the others knelt. That was deemed too distracting. Next day the Franklins were told to stand in the back row among the taller boys. Even so, their presence was judged discordant. The solution finally arrived at was that they come to school late, after prayers had finished.

A century later, Rosalind's school was more comfortable with such matters. While the rest of the school bowed their heads, the Jewish girls went off to a room by themselves – a practice laughingly called 'Jewish prayers' which gave them a chance to catch up on their homework. They joined the rest for the remainder of the school assembly and for the notices.

How many Jews there were at the school is unknown because no record was kept of religious affiliation. The Mercers' Company accepted the conscience clause of Gladstone's Endowed Schools Act of 1869 which entitled pupils to equal opportunities in education regardless of faith or lack of one. Indeed, in 1946, eight years after Rosalind had left St Paul's, Miss Strudwick proposed a quota 'to avoid too great a preponderance of Jewesses'. The Mercers flatly rejected the idea.

Rosalind was free of another problem that had plagued her forebears. Her father remained bitter about his own schooldays at Clifton College where he lived in 'the Jewish house' and, although an avid cricketer, was prevented by religious observance from playing sports on Saturday when all the matches were held. The same restriction had hobbled his sister Mamie at St Paul's where it was understood that Jewish girls could not be in the teams: 'the only real disadvantage I have felt in being Jewish in all my life', said Mamie.

Ellis was contemptuous of the over-literal orthodoxy which had forced his elderly father to be pushed in his bath chair and his elderly mother to struggle on foot to David's bar mitzvah in 1933 in order to avoid driving on a Saturday. His relaxed practice was fortunate for Rosalind who made the tennis, cricket and hockey teams at St Paul's. Photographs of successive hockey teams, dressed in pleated dark tunics, thick stockings, white shirts and neckties, show Rosalind maturing into a pretty, if somewhat guarded, girl with well-behaved dark hair held back by a velvet band.

'All her life Rosalind knew exactly where she was going,' according to her mother, 'and at the age of sixteen, she took science for her subject.' After the first four (pre-matriculation) years and achieving the General School Certificate, Rosalind entered the sixth form and was free to work towards the Higher School Certificate, thus concentrating at last on what really interested her: chemistry, physics and mathematics, pure and applied. She avoided the biology and botany courses taken by girls intending to go to medical school.

That she should have been so clear in her intention suggests that by the age of sixteen, if not sooner, Rosalind had realised what Albert Einstein had gradually learned about himself: that a scientist makes science 'the pivot of his emotional life, in order to find in this way the peace and security which he cannot find in the narrow whirlpool of personal experience'.

Einstein offered this self-analysis in 1918 at a celebration in Berlin for the award of a Nobel prize to Max Planck. He went on to describe the scientist's research as 'akin to that of the religious worshiper or the lover; the daily effort comes from no deliberate intention or program, but straight from the heart'.

Rosalind's science always came straight from the heart. St Paul's prided itself on its modern new science facilities, where Rosalind now spent much of her time. But was she well-prepared? When the Matriculation and School Examinations Council of the University of London paid a visit of inspection in 1935, it gave high praise to the 'magnificent new block of buildings given to science'. So too it admired the three 'highly-qualified mistresses' responsible for physics, chemistry and biology, and pronounced the mathematical teaching 'thoroughly sound and vigorous'. The committee nonetheless regretted that the older girls were not allowed greater access to laboratories in their free periods in order to get more experimental work done. The school's defence was that the extensive syllabus to be covered left the girls few free periods.

The examiners' report, combined with other observations including Rosalind's own, suggests that science was taught to girls in a different way than to boys: an intellectual endeavour calling for neatness, thoroughness and repetition rather than excitement and daring. The 'science' staff at St Paul's School for Girls included a 'gardening Mistress', for 'Nature Study and Gardening' which were parts of botany. Even at the higher levels of science, there was a great emphasis on order and the clear labelling of notes, diagrams and files. Anne Crawford had the impression that Rosalind did a lot of work on her own in order to get ahead more quickly.

One thing Rosalind did not learn in science or any other class at St Paul's were the facts of life. For those girls who did not do biology, the school offered the option of eight illustrated talks on general physiology and development, beginning with the amoeba, moving on to fish but stopping short of mammals. Rosalind did not take this course, nor did she get basic information at home. The Franklins simply did not talk about such things. In the wider family circle, it was joked that two of the three days of the Ellis Franklins' weekend honeymoon in 1917 were spent with Ellis trying to explain to his bride 'the meaning of marriage'.

An uncomfortable occasion in Rosalind's adolescence was the rare event, a dance with the St Paul's Boys' School. She and Jean Kerslake went together, with a boy summoned to share their taxi, an uncomfortable ride during which the trio sat in silence. To Rosalind's and Jean's dismay, they saw that all the other girls were wearing floor-length evening dresses. The two of them, in short taffeta party frocks, looked like children. No one asked either of them to dance the whole evening. Rosalind was sufficiently upset by the experience to report to her mother that her cousin Ursula had been wearing lipstick at the dance. Muriel, shocked, rebuked her sister-in-law, who retorted that she was delighted that Ursula should try make-up.

Rosalind had, Ursula felt, an 'incredible innocence' even for those prudish times, but an innocence shielded by a sharp tongue. When her school friends began to have crushes on their skating instructors or film stars, they would never talk about these with her. It was an inhibition she shared with her mother. Muriel was a highly intelligent, pretty woman, totally subservient to her genial, domineering husband. Mother and daughter got on well when they were alone together, as on a walking weekend when they explored the countryside around Chichester.

An object lesson in the messiness of affairs of the heart was delivered in December 1936 when King Edward VIII gave up his throne to marry the woman he loved. In January Ellis Franklin, an ardent monarchist, was heartily glad to see him go. Ellis had shepherded his family a few months earlier to watch the funeral procession of George V. Now he was scathing on the subject of the playboy king and Mrs Simpson. His wife had the benefit of his tirades: 'The weakness and lack of moral fibre of a member of the Royal Family who had not only broken up the second marriage of a dubious and unsavoury woman but was prepared to set her as his consort upon the throne of England!' Ellis was as relieved as any British subject when the uncrowned king abdicated in favour of his brother, a family man of irreproachable morals.

The whole Franklin family turned out for the Coronation on 12 May 1937. From Chartridge, Rosalind's grandfather arranged for all his family to have good seats along the procession route. In her thank-you letter, written on behalf of her brothers and sister and with her powers of verbal description at their best, Rosalind drew a vivid picture of an historic occasion and one of the happiest days of her life:

> Dear Grandpa,
> We have just arrived home, after a really wonderful day.
>
> We arrived at the club at 5.30 a.m., the trains were full but we got there quite easily. We went straight to our seats, and were both amazed at the wonderful view we had. We were in the front row, and had a clear view of the whole length of St James's Street, from St James's Palace to Piccadilly, while sitting in our seats. The windows have been taken right out, and the stands built slightly out into the roads, so, being on the left of the route, were just as near as we could be to the procession, and saw the whole thing perfectly, including the insides of the coaches.
>
> We had an excellent buffet breakfast at 7.30 and a six-course champaign lunch at 1 o'clock! We listened to the Abbey broadcast in the morning, and heard it very well indeed.
>
> By 6.30 a.m. the streets were as full as they could be. It was a very jolly crowd, and passed most of the time in community singing.
>
> At about mid-day, army lorries came round with rations for the men, each man being given his lunch in a paper bag. It was most amusing to watch them all feeding out of their bags, and to see the excitement of those who were served last when the final lorry appeared.
>
> It was fine the whole day until the procession was passing, and it was a great deal better that the rain should come then than earlier, drenching the crowds.

After the procession we were provided with an entertainment which I had certainly not anticipated – saw on a television set the whole procession passing Hyde Park Corner.

Owing to the procession starting earlier than was expected, we missed, while finishing lunch, the contingents of the Dominions and Colonies, and afterwards had cause to be very pleased that we had missed them, for being anxious to see them, we set off to look for them afterwards, and this led to our seeing a great deal more besides. As soon as we had seen the end of the television, and the crowds in St James's Street had cleared a reasonable amount, we set off towards the Mall. We were not allowed straight through, so we raced along Pall Mall, by Trafalgar Square, and through the mud in St James's Park, and arrived just in time to catch a glimpse over the heads of the crowd, of the royal coach arriving, and some of the guards.

We then went round behind the barracks, where we had been told we might find the colonial and dominion contingents. However, we saw nothing more than masses of guards and police, and gave up the search. Just as we were getting back to Buckingham Palace we saw the whole lot coming towards us, and they passed just in front of us. By that time a tremendous cheering crowd had assembled outside the Palace, so we decided to join them. We managed to obtain a very good central position on the steps of the Victoria Memorial and joined in the cheers and cries of 'We want the King' and 'For he's a jolly good fellow.' At last the door of the balcony opened – the crowd surged forward, waving handkerchiefs, gloves and hats, and a terrific cheer went up. It was a wonderful moment the King, Queen, 2 princesses, Queen Mary, Dukes and Duchesses of Gloucester and Kent, all came out. The crowd was solid right across the Mall and as far back as you could see. They stayed out for quite a time wearing crowns, etc., and immediately they went in the crowd sang God Save the King, and then dispersed.

We walked home, among huge crowds as there were impossible queues at all the stations.

If it does not rain too much more, we are going out to see the flood-lighting and the crowds tonight.

I really cannot remember a day that I have enjoyed more. Thank you very much for the marvellous opportunity you gave us.

Love from
Rosalind.

That she should have expended such effort at the end of an exhausting day that began at dawn is a testimony to her stamina and her sense of family obligation.

Uncle Herbert Samuel had an even better seat at the Coronation. He and his wife (Ellis Franklin's aunt Beatrice) sat robed in Westminster Abbey. Samuel had been awarded a viscountcy in the Coronation Honours List. Rejecting his first thought for his title – Lord Paddington – he chose instead the title of 'Samuel of Mount Carmel and of Toxteth in the City of Liverpool' with the motto 'Turn not aside', from 1 Samuel 6:12. It was a motto Rosalind might well have adopted for herself.

That summer Ellis and Muriel took their four oldest children to the mountains of western Norway. Roped together, all six followed a guide over the Josterdal glacier. Once again Rosalind served as family *rapporteuse*: sending precise descriptions to her grandfather of the twelfth-century cathedral of Bergen, the wooden houses, Ellis in his knickerbockers drawing sniggers in the streets; then, in glowing but precise terms, the glory of sunsets in deep purple mountains.

Her love of walking was acquired from her parents. So were the value of an arduous hike, the exhilaration of rocky peaks and the importance of scrupulous advance planning. The Franklins as a family were utterly undemonstrative – Rosalind may have kissed her mother on occasion, never her father or her brothers – yet to a rare degree, they spent their free time together, the children uncomplaining even as they got older. In Colin's recollection, 'we enjoyed plenty of fun and nonsense in family holidays, expeditions, country walks; my father was witty, with a fine sense of the ridiculous though that would never have applied to any aspect of his own life'.

Rosalind's last years at school were darkened by Hitler's rise to power, and its reverberations in London. In October 1936 Oswald Mosley was inspired to lead the black-shirted British Union of Fascists on a march through the Jewish neighbourhoods in the East End of London. The resulting Battle of Cable Street forced Mosley to abandon the march as police could not clear a way through the crowd.

After the Anschluss in March 1938, when the Austrian National Socialists invited the Germans into their country, the trickle of Jewish refugees turned into a flood. The *Jewish Chronicle* and the personal columns of *The Times* were full of pleading messages from desperate Jews seeking jobs as domestics, private secretaries or dressmakers so that they could get visas to enter Britain. In June 1938 after a visit from the Gestapo, Sigmund Freud and much of his family arrived in London. Freud was allowed into Britain easily as a distinguished person of international repute. However, for ordinary middle-class European Jews, at a time of British mass unemployment, entry permits were not readily granted. Rosalind now witnessed the total unanimity of purpose between her parents as they threw themselves into refugee relief work. Ellis reduced his time at Keyser's bank and the Working Men's College in order to help the Home Office organise the allocation of entry permits. He worked with Wilfred Eady (whose name was later attached to the levy imposed on cinema tickets to subsidise British films), then in charge of the Aliens Department.

Applicants were divided into those who could look to relatives or friends in Britain to support them and those who arrived entirely without support. Ellis was in charge of the 'guarantee department' of the German/Jewish Refugee Committee, with an office in Woburn Square. Ellis and his sister Mamie also set up an organisation to find homes for orphaned and homeless German Jewish children.

School friends of Rosalind's who previously had been enlisted to make sandwiches for the Working Men's College sports day now found themselves drawn into refugee work as the High Mistress of St Paul's released girls every day to help at Woburn House with the filing.

The Franklins offered more than administrative help. With 10,000 Austrian children arriving in London, many unaccompanied on the Kindertransport, they took two to live in their home as part of their family. Evi Eisenstadter, a young girl whose father was in Buchenwald, arrived in the summer of 1938 and shared a room with Jenifer on the top floor. 'Father Franklin', as she soon came to call Ellis, spoke to her in German.

A nine-year-old who had crossed half of Europe to get to London, Evi arrived on a Friday evening to be greeted with the information that she was being taken to the country for the weekend. The train was late, and when Wilson the chauffeur delivered her to the Franklin estate at Chartridge, the first sight she saw was two men coming down a grand staircase in evening clothes. They were so elegant, she recalled (giggling at the memory), 'I thought it was the King.' But it was only Grandpa Franklin and Viscount Samuel on their way to dinner.

Rosalind was fed up with school. Five months short of her eighteenth birthday, she went up to Cambridge to sit the entrance examinations in physics and chemistry. As usual, she feared she had made a mess of the papers and spent a tense few days during which she saw other nervous girls – 'nearly all north country' – trembling as they looked to see if their names were posted for interview. Rosalind did have one good interview in physics but did not enjoy a bleak halting conversation with the principal of Newnham who pointed out that if she began in October rather than spending a further year at school, in the so-called eighth form, she would be younger than the other new girls. As it turned out, however, both women's colleges, Girton and Newnham, offered her places. After ringing everybody she knew to get opinions of the two colleges she chose Newnham, to start that autumn, even though the college recommended that she wait a year as after another year's preparation, she would have a 'reasonable chance of an award'. But she did not want to wait, and told her grandfather, 'I am very much looking forward to going there.'

St Paul's was dismayed. An Oxford or Cambridge award brought honour to the student's school, with winners announced under 'University News' in *The Times*. Jean Kerslake did stay on at school for the extra year and won a Newnham scholarship because, her father told her, Rosalind had taught her how to work.

Family duties and social responsibility combined as the Franklins all worked together to help the refugees crowding into London from Germany and the recently annexed Austria. Grandpa Franklin, now a widower, being looked after at Chartridge by his unmarried daughter Alice, invited groups of émigrés to tea and asked his grandchildren to help him to entertain them. Rosalind willingly went down to Buckinghamshire to assist, as did her two siblings not at boarding school, Jenifer and Roland.

With school pressure off, she filled her time with extra-curricular interests, such as attending, under the auspices of her Aunt Mamie, who was now a member of the London County Council, a meeting of the Hammersmith Borough Council. She also attended an air raid demonstration in Kensington, the Chelsea Flower Show and a school debate on the motion, 'A return to secret diplomacy would further the cause of world peace.' Now considering herself grown-up, she was self-conscious about looking younger than her age. When one day on the bus the conductor asked if she wanted a 'whole or a half' – half fares were for the under-fourteens – she went and had her hair permed.

On 1 June she went off to France for her first visit there since her childhood trip to Menton and, now well-grounded in written French, she loved it. She joked that she was being sent to finishing school but threw herself into improving her conversational French even while disliking hearing her English friends speak with 'French mannerisms'. She learned her way around Paris, indulged herself in lunch at a pâtisserie, did 'masses' of sewing under the tutelage of a local woman, bathed in the Marne and bought dress material for frocks to be made up when she got home. Politely she asked her parents for 'some of my own money' to buy French books and to tip the maids.

While she was away, her mother received a letter from the High Mistress of St Paul's:

> Dear Mrs. Franklin,
> You will, I know, be glad to hear that Rosalind has won a School Leaving Exhibition of £30 a year for three years.
> Will you congratulate her for me and give her my best wishes.
> Yours sincerely,
> Ethel Strudwick.

The decision was not the school's but that of external examiners who assessed the records of the leaving class. They recommended Rosalind as one of four deserving of an award for outstanding performance: 'Rosalind Franklin showed great promise, and her work, especially in physics, revealed sound knowledge and genuine appreciation of the finer points of the subject.' The annual thirty pounds was a considerable contribution towards putting a student through university. Once more, her father would not dream of letting Rosalind accept the actual money and earmarked it for a refugee student. To be one of the best of the best: the honour was distinction enough.

The family was all the more pleased when Grandpa Franklin learned privately from their influential friend, General Maurice, former principal of the Working Men's College and a governor of St Paul's, that the Exhibition had been awarded because Rosalind had come first in the Cambridge examination in chemistry.

* * *

On 30 September Neville Chamberlain flew in to Croydon from Munich with his piece of paper. The new King and Queen were so happy with the promise of 'peace in our time', that they invited Chamberlain onto the balcony of Buckingham Palace to celebrate his triumph. (With half the country divided over Chamberlain's actions, this endorsement by the monarch of a controversial political act, was, according to the historian Andrew Roberts, 'the most unconstitutional act of the century'.)

The Munich pact looked different at the time. Viscount Samuel, no admirer of Chamberlain, thoroughly approved. The deal with Hitler, Samuel believed (and maintained long after 'Munich' became shorthand for selling-out), bought time for an unprotected British populace to prepare their defences against attack from the air.

And prepare they did. As Rosalind left London for Cambridge, trenches were being dug in the royal parks, public bomb shelters organised, and householders issued with sections of steel from which to assemble their own Anderson shelters (named after Sir John Anderson, the popular minister for ARP). A fifty-mile journey from London delivered her to a self-preoccupied university city where the gathering storm was scarcely noticed. Rosalind's main worry was the work that lay ahead. The Exhibition scholarship was gratifying, but even so, she asked her parents, 'Does this really mean I did the best chemistry papers in February?'

FOUR

Never Surrender

(October 1938–July 1941)

THE MYTH HAS GOT INTO potted biographies of Rosalind Franklin that her father opposed her going to Cambridge. That is not true. He would not have sent her, or later, his daughter Jenifer, to St Paul's if he had wanted a mere finishing school for his girls. Ideally Ellis Franklin may have preferred his elder daughter to be like his wife, or his mother Caroline Jacob, with her diploma from Bedford College and her service on the Buckinghamshire Education Committee. But he had long been aware that his elder daughter was nothing like them, or even like his assertive sisters who made their mark in public life but stopped short of gaining professional qualifications. But as Rosalind entered university, Ellis wanted her to do well as he did all his children. Moreover, he expected to be kept informed – weekly, at very least – of their progress and wrote them in return as his part of the dialogue.

The sound of the paternal voice, heavy with irony, comes through a letter to his second son, Colin, still at his public school, Oundle when Rosalind went up to Cambridge:

> . . . this is the first occasion on which you have told me your place in form without being pressed for it. It is of course a mere coincidence that you happen to be top on such an occasion. I congratulate you on the new standard that you have set yourself, and assume that you will be top each week in future.

Cambridge had admitted women since 1869, and Jews since 1871, but unlike Oxford, which had granted women degrees since 1921, it refused to accept them as 'members of the University'. Nor were the female students considered undergraduates, merely 'students of Girton and Newnham Colleges'. They were not entitled to the degree of BA Cantab., or to any degree at all, but rather to 'decrees titular'. The 'decree tit' made a good joke. Female students were admitted to men's lectures but, at least until the early 1930s, were expected to sit together in the front rows. If they came in late, they were liable to be pelted with paper or greeted with stomping feet.

They were not offended by these restrictions. To the contrary, Newnham and Girton students considered themselves lucky to be among the chosen 500 (the quota set for women so that their numbers would not exceed 10 per cent of the male undergraduate body). They giggled to hear themselves addressed collectively as 'Gentlemen'. They were delighted not to have to wear academic gowns, which, short as these were, might have got caught in their bicycle spokes, and to be free of the proctorial discipline enforced on the men. And they were all studying for honours degrees.

Newnham, founded in 1871, was within easy walking distance of the Cam, the city centre and the splendour of King's College. Its students were not isolated from male company. They had men as supervisors and often as research partners. Most of the university societies, and all lectures, were open to women and marriage was no bar to teaching. The second principal of Newnham, Mrs Eleanor Sidgwick who held the post from 1892 to 1910, lived in college with her husband Henry, philosopher and college founder.

Women were nonetheless anomalies in a medieval institution to which the monastic tradition still clung. Even those of high rank had no say in the affairs of the university. The mistress of Girton and the principal of Newnham were not allowed to participate in university ceremonies and functions but were required instead to sit, in hat and gloves, with the wives of the faculty at the ritual occasions when the men wore their scarlet academic robes and black velvet doctors' hats.

Few first-year university students could match Rosalind's diligence in keeping in touch with home. The 'Dear Mummy and Daddy' of her schoolgirl years had long been replaced by 'Dear Mother and Father', but she shared with them the minutiae of her new existence, from studies to clubs to the Venetian print hung on the wall. She boasted of her small economies – buying a second-hand chair, a second-hand bicycle and used textbooks, and choosing life, rather than annual, membership of the Chemical Society because it saved 6 shillings over three years – or 7 shillings, if by any chance she should stay for four years. (This was the first hint to her parents that she might stay on at Cambridge for postgraduate work). She was also:

> being dragged into the Jewish Society, which is very expensive (30 shillings, most others are 3/) and no use to me as I have neither Friday evenings nor Saturday mornings free, but I shall have to join as Grandpa has been writing to the professor in charge, and I have been asked to lunch there on Sunday. I went to a social evening of the society at a cafe last night – they are an awful crowd of people.

Newnham was in many ways like boarding school. 'This evening there is an awful thing called the college feast in which the whole college dines together in a great crush in evening dress, and afterwards scholars etc. are solemnly "sworn in" in private.' Rosalind groaned at having to participate in a fresher's play in which her role was to walk across the stage on an evening when she had wanted to go to the Cambridge Union to hear talks by the geneticist J.B.S. Haldane and the chemist Alfred Noyes. Seeing that other girls had brought their gas masks with them, she asked to be sent hers but scoffed that the college was making 'a ridiculous fuss about ARP'.

Rosalind was not one to put up with silly rules in silence. An early victory was winning the principal's permission to be away from college overnight to attend the Founders' Night dinner at the Working Men's College in London. The principal 'happened to know about the Working Men's College and said she would regard it for me as a sort of "family function" – apparently the only acceptable occasion for a night. She had already refused nights for other people.'

In Cambridge as in London, Rosalind found herself surrounded by relations or those claiming to be so. Attending an engagement party for a refugee couple, she was hardly in the door when the hostess, a woman from Breslau, said 'tell me all your genealogy. I want to know if you are connected with someone called Ellis Franklin.' At a tea party a Roger Hartog appeared and claimed to be a third cousin. And 'Who and what is Dr Redcliffe Salomone? I know him by sight, but that is all. I have been asked, through a 3rd person, to go to lunch with him some time.'

Rosalind threw herself into her courses: chemistry, physics, mathematics and, having bought some good instruments, mineralogy. She signed up for extra chemistry and a course in scientific German. She had come to the right place.

Cambridge had been pre-eminent in mathematics since 1669 when Isaac Newton, at twenty-seven, became Lucasian Professor of Mathematics. In 1851 Natural Sciences were added to the tripos (honours examinations named after the three-legged stool on which eighteenth-century students sat to be examined). In 1871 the Cavendish Laboratory was founded, with James Clerk Maxwell, who unified the theories of electricity and magnetism, as its first professor of experimental physics. In 1897 at the Cavendish, J.J. Thomson discovered the electron, opening the way for the development of modern physics.

Plunging into work at university level, Rosalind had the common freshman experience of thinking that everybody else was better prepared. Overwhelmed, trying desperately to keep up with the reading in order to understand the laboratory assignments (the 'practicals'), she decided she had been poorly taught at St Paul's, especially in laboratory techniques. The elegant new science block at her old school now seemed 'to be all show and nothing in it'. She was very glad she had not wasted another year at school.

At the same time she welcomed the stimulation of the university environment. She joined the Archimedeans, a mathematics society, and went to 'a most exciting' lecture on the theory of fluorescence, and another on penguins and whales. She listened to the prevailing great names of Cambridge science, including J.J. Thomson and J.B.S. Haldane, whose mathematics she couldn't follow, and went to a meeting of the Association of Scientific Workers 'over which Prof. Bragg is presiding'.

This occasion gave her a look at the father (with an assist from his own father) of X-ray crystallography. In 1915 William Lawrence Bragg, at twenty-five, had shared a Nobel prize with his father William Henry Bragg for demonstrating the use of X-rays for revealing the structure of crystals.

Bragg's Law, named after Bragg junior, not senior, builds on the fact that a crystal by its nature suggests an orderly pattern of atoms inside. When X-rays are shone through it, the atoms diffract – that is, scatter in particular directions – and leave spots on a photographic plate. Bragg's equation relates the positions of the spacing of the atoms and thus to the structure of the molecules that make up the crystal.

Rosalind was eager for such knowledge and technique (which would form the basis for her entire body of professional work). As an undergraduate she knew she was beginning at the beginning. In her new notebook which she headed 'Mineralogy', she wrote at the top of the first page, 'What is a crystal?'

Had she asked herself 'What is a crystallographer?', she could have profited from talking with Bragg's son Stephen. Assessing his distinguished father, he judged: 'By and large, people who choose to go into science are not greatly interested in psychological problems . . . He liked the exactness of science, and the perfection of its truths – humanity could be rather messy in comparison.'

Metaphors of combat filled Rosalind's letters just as they filled the daily newspapers. By January 1939 she could report a victory.

> The complaint about my chemistry lectures was well worth making – I now go to much better ones . . . also, to better physics and chemistry labs. I am in the midst of a struggle over maths lectures, but I'm very much afraid I'm on the losing side. The ones I want to go to (I have been to two of them) are on analysis. I have heard from several people who have done the course that they are very interesting and also provide useful short cut methods in physical calculations. The lecturer is very good, though female.

The lecturer judged 'good, though female' fared better than the lecturer on solar prominences, whom Rosalind pronounced 'not much of a scientist. He ranted when he discussed phenomena which have not been explained – but that may only be because he was American.'

'Good, though female' was no small praise in a university that in more than 700 years had had no woman professor. Rosalind witnessed the appointment of the first, the archaeologist Dorothy Garrod of Newnham, and laughed at the embarrassment caused among the men's colleges. The Cambridge custom was to invite newly elected professors to a 'feast' at each college, but women were not allowed at feasts. What to do about Professor Garrod? In this case, as Rosalind reported home, King's College broke with tradition and invited Garrod to a ceremonial dinner. The other colleges held back, and Newnham had its own feast.

For relaxation, Rosalind threw herself into strenuous sport, switching from hockey to squash in her second term. She went boating on the Cam, played endless sets of tennis, went on thirty-mile bicycle rides over the Gog Magog hills, and when the term was over, rode back to London. 'Why,' she demanded of her startled parents, 'are you so surprised about cycling home? I want my bike in London, and it seems the simplest way of getting it there . . . It isn't very far.'

Working long hours in the lab, she made few close friends at university. Girls who lived in the adjoining rooms thought of her as quiet, and keeping to herself. She preferred her old St Paul's friends to the new girls she was meeting and stayed aloof from the social life of the college.

One who broke the barrier of this reserve was Peggy Clark from Bristol, in the same year at Newnham and taking physics and maths with the same lecturers. As their friendship developed, another Newnham student said to Peggy, 'I don't know what you see in Ros – you know she is a Jew, don't you?' Peggy was staggered because she had no personal knowledge of anti-semitism. When the flooded fens made excellent skating rinks and Peggy could not afford to buy skates, Rosalind lent her hers, skating by herself in the afternoon so that Peggy could use them at dusk. The two protested jointly about inefficient physics tutorials and a maths lecturer who wrote in such small print on the blackboard that nobody could read it.

The enemy, to Rosalind in 1938, was fascism and, its ally, pacifism. So opposed were some girls in her college to any form of preparation for war that they refused to carry gas masks or take part in any air raid drill. A lecture given by Lawrence Housman on 'The Price of Peace' infuriated Rosalind. 'I would not have gone if I had known he was pacifist – it was awful,' she wrote home. 'The majority of people, however, believed all he said and came away converted. I have been busy to-day unconverting them, and had several successes.' Inclined to the political left but never the far left, she steered a course between the pacifists, who argued that war with Germany was unnecessary, and the Communists. She decided against joining the Socialist Society because of 'a horrible young man' who talked of "Socialists and Communists" as one'.

Her father shuddered to hear these sentiments. Ellis was a free-marketeer and the most conservative of his family. Rosalind enjoyed stirring him up, knowing that on the issue that concerned them most – the plight of European Jews and the need to stand up to Hitler – they thought as one. In the weeks following Kristallnacht (9–10 November 1938), when Jewish shop windows were smashed, many Jews were killed and thousands dragged off to concentration camps, Rosalind was deeply upset at Cambridge's indifference. 'Apart from your letters and *The Times*,' she wrote home, 'I would still have no idea that anybody objected to Germany's treatment of the Jews. People do not talk politics here, but I still hear them discussing the colonial question in the same way as before.' A Cambridge Union debate that month tackled the proposition that 'The continued existence of the British Empire is a danger to World Peace.'

Rosalind was disgusted too by the Home Office's reluctance to admit refugees without sponsorship or means of support, which (she felt) accordingly let in only a small number compared to those being admitted to the United States. The Board of Deputies of British Jews appealed to the government to give larger opportunities for resettlement elsewhere in the Empire and in Palestine. However, the Board, fearing a resurgence of British anti-semitism, which it suspected was endemic in the Foreign Office, instructed those refugees who did arrive not to make themselves conspicuous, talk in a loud voice or engage in politics.

Caution was advisable. The young American historian Arthur Schlesinger, then a student at Cambridge, stood in Piccadilly Circus and watched the Blackshirts surging through, shouting 'Mosley! Mosley!' and 'Down with the Jews!' and a crowd of young men beating up people 'of Jewish appearance'.

In November Neville Chamberlain issued a statement to the effect that, regrettably, the number of refugees Britain could admit was limited by the capacity of the voluntary organisations to receive them. As for resettling them in 'the Colonies and Protectorates and Mandated Territories', His Majesty's Government did not wish to prejudice the interests of native populations there by permitting unlimited immigration. Nor could the government ignore the fact that many of these areas were 'unsuitable either climatically or economically for European settlement'. The statement effectively limited Jewish immigration to Palestine to 75,000 within the next five years and forbade the selling of land to the Jews. It was a harsh blow to Zionist aspirations.

Rosalind felt isolated. 'I cannot see why there has not been more criticism of the inadequacy of the Prime Minister's "statement",' she wrote home. 'Why is France so indifferent?' She joined a group in Cambridge trying to raise money for refugee aid. But Cambridge was the last place to look for militancy. In the Amateur Dramatic Club's Christmas show in 1938 a joke about 'one man's small moustache' was taken out (at the Lord Chancellor's insistence) for fear of causing offence to Herr Hitler.

A reminder to the Franklins of the obligations of faith came when Arthur Ellis Franklin – 'Grandpa' – died in January 1939 at Chartridge. He left the estate to his eldest son, Cecil, who sold it to a London firm seeking a country refuge during the war. Only the gardener's cottage at Chartridge, with its five bedrooms, plus a two-bedroom flat over the garage, were kept for the family's use. Rosalind, like the other grandchildren, inherited £100 immediately.

Grandpa Franklin had had two conspicuous failures in his attempt to persuade his children to marry within the faith, as his own father had succeeded in doing with all seven of his. Arthur's eldest son, known to the family as Jack, had married an Agnes Foley at a registry office. His radical second son, Hugh (of the Churchill dogwhip incident) had twice married non-Jewish women named Elsie. The first (from whom Rosalind got her middle name) converted to Judaism before dying of the Spanish flu in 1919. The second Elsie did not convert. His father cut Hugh off with a pittance and never saw him again.

Making his will, Arthur Franklin sought to ensure that none of his eight grandchildren would make the same mistake. In a lengthy, tortuously worded legal clause, he proclaimed:

> WHEREAS my forefathers have been steadfastly loyal to the Jewish Faith and I have to the utmost of my ability followed in their footsteps . . . I EARNESTLY REQUEST my named children to uphold these principles and to inculcate them in their children and especially I trust that no descendant of mine will intermarry with a person not of the Jewish Faith or renounce Judaism or the Jewish Faith.

Any descendant who married a non-Jew would be treated as having died unmarried within Arthur Franklin's lifetime – that is, with no entitlement to any income from capital or property. Moreover, none of his estate was to go 'to any person born out of wedlock' or 'strangers in blood to myself though adopted by a member of my family'. (Many years later, such legal ingenuity was struck down by a court ruling against testamentary attempts to rule lives from beyond the grave.)

This wording had no practical meaning for Rosalind or her siblings, not having finished education nor entertained thoughts of marriage. But the constraints within or against which they would make their future choices were clear.

By spring Rosalind's thoughts had turned to the holiday she wanted to take before the May examinations – for which she felt desperately behind in her preparations, especially in physics and chemistry. She wanted to go hostelling in the Peak District, with her brother David and her second cousin Catherine Joseph (their old St Paul's feud long forgotten). Months earlier, she had sought parental permission: 'Please don't just say "no" . . . I know lots of people who have been on these NUS [National Union of Students] trips and they have all been very successful . . . I am sure you will say next "how are you going to pay for it?"' In anticipation of this objection, she itemised various savings, gifts and receipts which gave her nearly £50 in hand (not to mention her recent inheritance). 'So you see I could almost pay for the whole family to go (this does not mean I intend to).'

Planning the trip was a pleasure in itself. She was proud of her equipment – her hiking boots, her field glasses and guidebook collection – as well as of her skill at route planning and reading timetables with an eye out for the cheapest or most unusual route. She also felt glee in pointing out (in her mother's tart memory) to the travel clerk that the information given her was wrong.

The holiday was a temporary respite from the apprehension about exams. She still lacked confidence. However, she was not shaken by her supervisor's report from Newnham, which included two comments: 'she knows her work but does not always keep to the point' and 'she does not seem to take criticism kindly'. Both meant – Rosalind interpreted for her parents – 'she doesn't agree with me'.

The exam ordeal, when it came, was worse than she had expected. Her confessional letter of 20 May 1939 makes painful reading for anyone aware of the female capacity for self-doubt:

> I have made a frightful mess of exams . . . fairly easy papers which I should have done really well. Mineralogy, which I expected to do best, was a lovely paper but I wasted the first ¾ hr completely – going about a question the wrong way, and having to do it all again later. The result was a terrific rush and a lot of bad blunders. I think I did best in maths – the one subject which doesn't matter . . . I have never done anything so badly as yesterday's physics practical. I did ½ instead of 2 experiments – and I don't think I can possibly get more than a third in physics now.

Rosalind was wrong. She did not get a third, she got a first in these preliminary examinations. Her total marks, for physics, chemistry, mineralogy and maths, put her in joint second place.

The way was now clear to sit part one of the Natural Science tripos the following year. You could spend three years preparing for part two, she explained to her parents, 'if you are less intelligent'. She was highly motivated to do it in two. Only those women who got a first in part one of their tripos were allowed to go on to part two, which was considered the more important and interesting part of the university course.

As she turned nineteen in July 1939, Rosalind still found it important to maintain the attention of Ellis and Muriel. When her parents not only visited but sent four letters and a cake, she joked, 'Now I really feel I am being treated as I should be, and can forgive you for all the letters you have ever not written to me.'

That summer the Franklins once again went to Norway *en famille* – taking all five children this time, and Nannie too, who arrived later, with Jenifer who was recovering from chickenpox. It was an ambitious undertaking when barrage balloons were floating over London, when there was talk of a possible German invasion and the children had already built their own air raid shelter in the garden of '5 PP'. But the Franklins crossed the North Sea anyway, for Ellis reasoned that if war broke out Jenifer would not have a chance to go abroad for many years. So off they went and found their familiar guide to lead them through their old haunts for fishing and climbing. They were in Fjaerland, with David and Rosalind planning a trek across a glacier, when news of the Nazi-Soviet non-aggression pact of 24 August 1939 reached them. Ellis immediately bundled the family onto the ferry from Bergen to Newcastle. It was the next to last boat to leave.

As Britain waited for Hitler to strike, Rosalind returned to Newnham to begin her second year. Her father at first refused to pay for it, thinking she ought to do war work; David had left Oxford after two years to join the army. But Rosalind's mother and her Aunt Alice said they would pay instead and he relented. Cambridge was full of evacuated children and RAF servicemen, and there were queues outside all the shops. At the college, black shades had been put on bright lights because no weaker bulbs were available. To save gas, there was no toast for breakfast. Rosalind was put in charge of waking ten people when there was an air raid alert, the signal for the whole college to go into the trenches dug nearby. Fear of attack was not unreasonable as Cambridge was surrounded by airbases.

The raids never came. The 'phoney war', to hard-working Rosalind, was time utterly wasted. Irritated by the meaningless ritual and constant false alarms, she sought a confrontation with Mrs Palmer, college tutor for her hall of residence (Peile Hall). Once more she won – at the price of hearing some ugly words applied to her character:

> I have just had a great triumph, though a somewhat disagreeable one. I and a few others decided it was time something was done about the . . . going out to the trenches . . . for every warning – we have had only 1 quite undisturbed night – yesterday the thing went at 7.30 and lasted until 11.15. I *had* to do some work. We were the only college being handicapped in this way – and I was getting badly behind. So three of us stayed in on the first floor and boldly left the light showing through the door. Mrs Palmer came in in a storm and turned us out, saying we were 'disloyal, deceitful and untrustworthy' and were to see her today. Before we saw her she had called a meeting to say that nobody need go to the trenches before 11 PM!

Rosalind did have a lot of work to do. There was no less science to master just because there was a war on. In 'optics' she drew a great many diagrams of lenses and light passing through a slit, and traced the history of the subject back to Newton, Descartes and Doppler. In physics, she studied the first and the second law of thermodynamics, and read Linus Pauling's classic text, *The Nature of the Chemical Bond* (explaining how electrons hold molecules together). She learned about proteins that fold; about the infectious tobacco mosaic virus that could be extracted from a tobacco plant and crystallised in a bottle, and also about the nucleic acid contained in chromosomes. She noted the experimentally useful form of nucleic acid, sodium thymonucleate (obtained from calf thymus glands), with its high molecular weight of 800,000 (now known to be much greater) and its bases stacked up at 3.4 Ångströms along its chains. (Bases are the opposite of acids, chemical compounds that take up the ionised hydrogen produced by acids. In the nucleic acids each base is linked to a sugar and a phosphate group, making up a nucleotide.) A sketch in her workbook represents a helical structure. She made a note to herself: 'Geometrical basis for inheritance?'

Getting deeper into crystallography, which would become her expertise, she joined the small band of the human race for whom infinitesimal specks of matter are as real and solid as billiard balls. She easily met the first requirement of the profession, the ability to think in three dimensions. The 'Ångström' (named after the nineteenth-century Swedish physicist Anders J. Ångström) was now part of her working vocabulary as the unit for measuring extremely short lengths. One Ångström represents a hundred-millionth of a centimetre.

After Lawrence Bragg, crystallography was developed further at Cambridge in the 1930s by the brilliant and ebullient J.D. Bernal, who refined the nineteenth-century classification of 'space groups' – the 230 forms into which the seven recognised crystal systems are organised. Rosalind made notes on 'Methods of approaching structure from space group', which included the observation that 'A molecule which is long or flat may in general be entirely contained within a space having the same size and shape as the unit cell of the crystal.' She drew diagrams of all types, and noted: 'Monoclinic all face centred.'

She understood very well what is easily confused by an outsider – that the marks appearing on a photographic plate were *not* of the atoms themselves inside the crystal, but rather of the spots that X-rays make when scattered by hitting the atoms. The

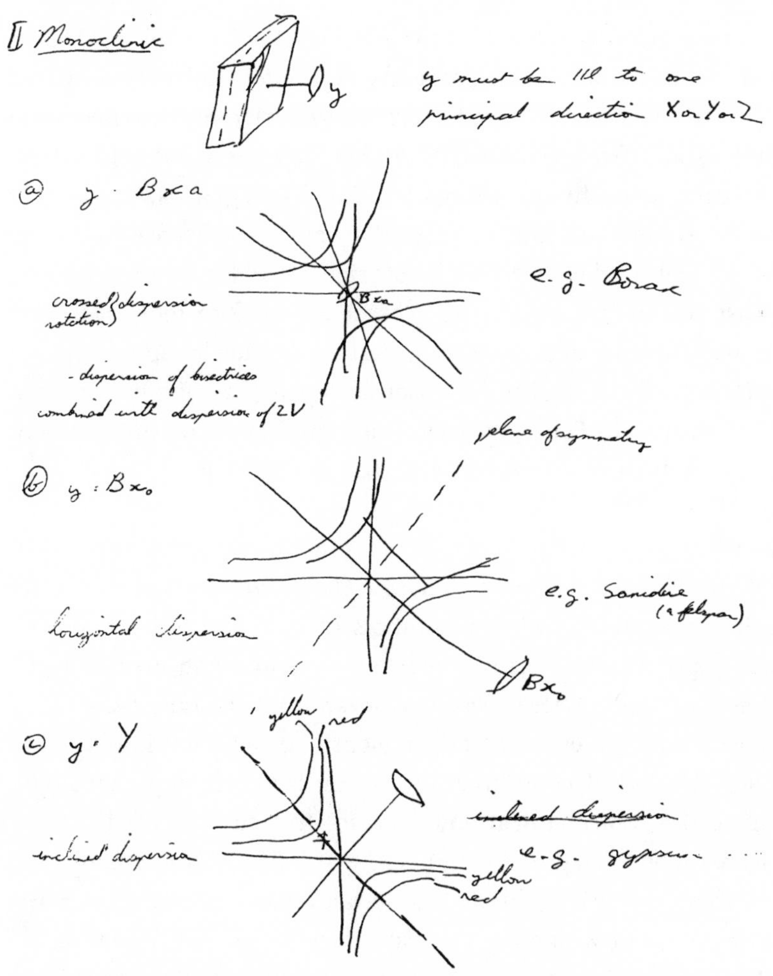

Rosalind's undergraduate notebook headed 'Crystal Physics' shows her learning the space groups and properties of various crystal forms.

spots vary in intensity as the X-rays reinforce each other in some directions and cancel each other out in others. From the position and the intensity of the spots, the atomic structure of the crystal may be guessed at. She learned also about the best angles from which to aim X-rays for efficient diffraction, and how to rotate the crystal to take photographs from many angles. 'Absorption of X-rays depends only on number and kind of atoms present,' she reminded herself in clear handwriting.

Away from the lab she experimented with cooking in her room, from the meagre ingredients available. One supper shared with a friend was composed of fried eggs, fried mushrooms, fried potatoes and fried crumpets. Another day she went into a fruit shop and saw '"white things looking like eggs" and was told they were and I could have four! I had lots of spare marg', so we had real fried eggs with bits of fried bread for Sunday supper, a feast which it is still quite a great pleasure to look back on.' She also dabbled in politics and worked on the election campaign for Dr John Ryle, Regius Professor of Physic (i.e., of medicine, not physics), who in February 1940 was standing as the Independent candidate for the university's parliamentary seat.

The war was pushing women to the centre of academic and industrial life. The proportion of women at the university had increased, as the total undergraduate numbers of men and women shrank from 5,491 men and 513 women in 1938 to 2,908 men and 497 women in 1940. There were Girton and Newnham graduates in almost every government department – in military intelligence, in signalling and code-breaking, in anthropology, archaeology, map reading, mechanical engineering. Indeed, one of the social features most sharply distinguishing Britain from Germany, the Newnham historian Gillian Sutherland has observed, was the participation of women in the war effort. It was not unreasonable of Rosalind to dream of a challenging job once she had completed her third year.

She followed closely the progress of the war, which by January 1940 had enveloped the Finns, who were first bombarded by the Soviet Union, then invaded. When a British naval ship near the coast of neutral Norway attacked the German prison ship the *Altmark* and freed nearly 300 British seamen prisoners, Rosalind declared to her parents that Britain had lost its moral advantage, giving enormous opportunities for enemy propaganda. Her father reproached her for such an unpatriotic opinion. She shot back: 'I am *not* one of the people who always says England is wrong.'

She thought it was almost as bad to say that England could never be wrong – a provocative opinion that she hoped did not mean 'that the reply will again have to occupy 2 pages of Father's letter and so have room for nothing else'.

By the spring of 1940 the war was phoney no longer. The Nazis invaded Denmark and Norway, then Holland. The Dutch surrender after six days provoked another father-daughter skirmish. Rosalind wrote Ellis:

> I don't understand your remarks about Holland. It doesn't seem 'weak-kneed' to give in after they had lost ¼ of their country, all their air force . . . and about to lose the rest and many civilians as well. You didn't say that about Finland and the Finnish losses in 3 mo[nth]s were less than the Dutch losses in a few days. It is obviously all we can do to save Belgium.

When Belgium surrendered, however, Rosalind was furious. She wondered if King Leopold had the constitutional right to do what he had done and wondered if 'our King' could do the same?

Her preparations for the all-important part one of the tripos coincided with the grimmest news of the war so far. German armies were poised to push down into France and turn west towards the English Channel. The fall of Paris was imminent and the British Expeditionary Force on the Continent had withdrawn to the beach at Dunkirk. On 10 May the House of Commons forced Chamberlain to resign and Winston Churchill became Prime Minister over a national government.

Rosalind disagreed with her father's faith that common sense and right would triumph. Common sense, she argued, would say that 'we are being beaten . . . Incidentally, America is not likely to let us lose though she seems willing to let us come very near it.'

Her personal outlook was also gloomy: 'Exams begin on Saturday. I wish I could manage to work 10 or 11 hours a day like most people do now but I can't.' She was cutting her laboratory sessions in order to revise and did nothing but work, sleep and play a little tennis in the long evenings created by 'Double Summertime'. As she got into the exams, she felt she hadn't 'got a scrap of brain left . . . I have never doubted that I shan't fall below a Second but I wanted a First and don't think there's any hope of that now.'

In ten days between the end of May and early June, over a third of a million British and Allied troops were evacuated from Dunkirk. On 4 June Churchill gave his 'We shall fight on the beaches . . . we shall never surrender' speech. On 19 June, as he broadcast the news of the fall of France, Rosalind was back in London, listening to the BBC while helping her mother sort out clothes for refugees at the Women's Volunteer Service in Eaton Square. Once again her pessimism had been unnecessary. She had got a first in part one of the Natural Science tripos, and won a college exhibition scholarship – £15 – for her final year.

If, as seemed possible, Cambridge might close down before the autumn term, Rosalind thought she might get work as a chemist. To have passed part one of the tripos was the equivalent of a degree. Her father suggested some form of 'land work' – agricultural labour – but she rejected it out of hand. She would be 'quite exceptionally bad' at anything except science. This flat declaration prompted Ellis to accuse her of being interested in nothing but science – in fact, of making science her religion. In a sense, he was right. Rosalind sent him an eloquent four-page declaration whose length testifies to his centrality in her life:

> You frequently state, and in your letter you imply, that I have developed a completely one-sided outlook and look at everything and think of everything in terms of science. Obviously my method of thought and reasoning is influenced by a scientific training – if that were not so my scientific training will have been a waste and a failure. But you look at science (or at least talk of it) as some sort of demoralising invention of man, something apart from real life, and

> which must be cautiously guarded and kept separate from everyday existence. But science and everyday life cannot and should not be separated. Science, for me, gives a partial explanation of life. In so far as it goes, it is based on fact, experience and experiment. Your theories are those which you and many other people find easiest and pleasantest to believe, but so far as I can see, they have no foundation other than they lead to a pleasanter view of life (and an exaggerated idea of our own importance) . . .
>
> I agree that faith is essential to success in life (success of any sort) but I do not accept your definition of faith, i.e. belief in life after death. In my view, all that is necessary for faith is the belief that by doing our best we shall come nearer to success and that success in our aims (the improvement of the lot of mankind, present and future) is worth attaining. Anyone able to believe in all that religion implies obviously must have such faith, but I maintain that faith in this world is perfectly possible without faith in another world . . .
>
> It has just occurred to me that you may raise the question of a creator. A creator of what? . . . I see no reason to believe that a creator of protoplasm or primeval matter, if such there be, has any reason to be interested in our insignificant race in a tiny corner of the universe, and still less in us, as still more insignificant individuals. Again, I see no reason why the belief that we are insignificant or fortuitous should lessen our faith – as I have defined it.

This humanistic credo comes perilously close to a renunciation of the Jewish faith as proscribed in her grandfather's will. But Jewishness is more than a religion. Rosalind was, says her sister Jenifer, 'always consciously a Jew'. If Jewishness is understood to mean unswerving loyalty to family, a belief in the importance of knowledge, especially in science and medicine, and the virtue of hard work – even over-work, Rosalind remained true to her tradition.

* * *

Holidays were not forgone, but rather taken in England. In August 1940 she went on a walking tour in the Pennines and the Lake District with another relative, this time her cousin Ursula Franklin. Rosalind was a far quicker and, Ursula felt, far braver, walker and could get quite impatient if she had to pause to wait. One night, with their backpacks, a heavy mist came down and there were no lights anywhere. They hadn't allowed for the blackout in their planning. Totally lost, they came to a road and walked for an hour without seeing a house. Suddenly there was a farm. Soaked to the skin, they knocked on the door and all was well. The farmer's wife put them up in an ancient room with a feather bed and gave them high tea, ham and eggs, for 7 shillings 6d. each. It was the kind of adventure Rosalind savoured.

On their tour, the cousins met a town clerk and his wife from the north of England. To Ursula's surprise, Rosalind paired off – in an utterly non-flirtatious way – with the married man. The two fell into rapt conversation and became such good friends that they wrote to each other afterwards. This phenomenon would occur frequently in the years to come: Rosalind enjoying the role of a non-threatening other woman in a triangle.

Rosalind's third year at Cambridge began during the Blitz, which started on 7 September 1940 when for four days 900 German aircraft pounded London and 1500 people died, mainly in the East End. The stoicism of inner-city Londoners spending their nights in the Underground was romanticised for American consumption as plucky English fortitude and Cockney cheeriness. Ellis and Muriel Franklin themselves took this view. (Their suggestion that the sleeping conditions in the Underground were probably more hygienic than the homes where many of these unfortunate people lived was another parental opinion which drew Rosalind's scorn.) When in November a bomb blasted out all the window glass at 5 Pembridge Place and killed two old women in a house in Chepstow Place at the rear, the Franklins decided it was time to leave and took a house in Radlett in

Hertfordshire. Nannie retired to Shropshire. Rosalind, from Cambridge, was concerned about her things being moved out of her room in her absence, particularly her desk:

> If it does not go, I would like *everything* out of it, in as little muddle as possible – I know exactly where everything in it is, though it may look confused . . . Also nearly everything from the drawer of my bookcase and my climbing boots from the cupboard underneath – I couldn't bear to have them bombed . . . As for books, I don't think there are any I can say I 'specially want', but I would like, naturally, to have as many with me as possible. One cannot live in a house permanently without books . . . in particular, I might mention all French books, my French dictionary, and encyclopaedia – though this does not mean that I don't want any others.

She sympathised with her mother having to set up a new household: 'I suppose "unfurnished" means you have to take carpets, curtains and everything, and none of them will fit.'

Rosalind entered her third and final undergraduate year at Cambridge, in October 1940, in a stronger position than she had expected. If the college were to close down, she had sufficient credentials, with her first in part one, to do war work as a chemist.

Cambridge was no longer ignoring the war. Scientists and other male faculty were vanishing into war research or, if Jewish refugees, confined in internment camps to forestall the feared 'fifth column'. 'Practically the whole of the Cavendish have disappeared,' Rosalind reported. 'Biochemistry was almost entirely run by Germans and may not survive.' Among the internees taken was the young Vienna-born Max Perutz, who had been a rising star in crystallography at the Cavendish since 1936.

To prepare for part two of the tripos, she recruited a new supervisor for herself. She chose Fred (later Lord) Dainton because she was now specialising in his field, physical chemistry – the blend of the two disciplines of chemistry and physics exploring the structural characteristics and behaviour of atoms and molecules.

Dainton, who had a full schedule and did not want to take her on, succumbed when Rosalind presented herself at his door in October and declared, 'I want to have you as my part two supervisor. I don't think Delia Simpson [her previous supervisor] entirely approves. Will you take me?' Dainton, a Yorkshireman, was attracted by her directness – not what he associated with 'middle-class Londoners'. He liked her willingness to share the hour with another student to whom she was very helpful. As the year wore on, he felt he was over-working her, setting her a long essay each week and insisting that she defend her point of view: 'something she was never loath to do'. Harder to persuade her to relax; he found her writing and her manner 'a little crabbed'; she was, he thought, 'a rather private person, with very high personal and scientific standards, and uncompromisingly honest'.

Rosalind did not choose to relax by joining the social whirl. A quarter of a century before the sexual revolution, the young men and women students of Cambridge went dancing, fell in love, had affairs, became engaged. There were belles of the university who were there for a good time and to catch a man. College rules acknowledged human nature to the extent of insisting that men had to be out of women's rooms by 7 p.m. Rosalind's St Paul's friend Jean Kerslake, caught up in parties and dates, felt that Rosalind believed she was flighty and disapproved of her not working harder.

Even so there were others, such as her friend Peggy Clark, who saw themselves, while not opposed to marriage, as wanting to be independent, to prepare to earn their own living and to make use of their academic training. It was Rosalind's personal choice to stay out of *la ronde*. When, for whatever reason, she was bold enough to invite a cousin, Stephen Waley, to Radlett for the weekend after Christmas, she was so afraid that the young man might try to hold her hand that she bribed her fourteen-year-old brother Roly with a sixpence to walk with them.

Despite her wariness of sexual attraction, she took great care with her appearance. Although not conventionally pretty, she had a trim figure and intense dark eyes set off by a pronounced widow's peak. Even as an undergraduate she showed a flair for understated sophisticated elegance in her clothes and seldom made a mistake. For a commemoration dinner at the college, she sent home for a very precise selection from her wardrobe:

> please send my evening dress (tulip one), evening shoes and evening petticoat. Shoes in bottom drawer of the wardrobe (gold *or* silver). As for suggestions for a gift, P.S. I forgot to answer the most important part of your letter I should very much like a handbag – my present one is the first and only one I have ever possessed and is falling to pieces inside and out I should like it black, because my present one is brown and I have to leave it behind when I wear my one and only silk frock which is black.

In the first term of her final year, Rosalind encountered, like an apparition sent from on high, a woman entirely different from any she had known. Adrienne Weill was a French-Jewish scientist: commanding, handsome, inspirational, intellectual – and also a widowed mother brave and shrewd enough to get herself and her daughter out of France in response to de Gaulle's call for French people to join him in England.

Her own mother's life seemed impoverished in contrast. In a letter to Muriel, who was now organising her temporary country home, Rosalind said:

> I can't think why you spend so much of your time cooking and washing up and thinking about nothing else. You have got Nannie Alice and a daily . . . and though they might not be quite as efficient as ordinary maids you couldn't possibly have more than two in a house that size. When I said all of this to you at home . . . You used to [have] a huge house of many floors, many stairs, numerous rooms, a large and inconvenient basement kitchen and pantry miles from

> the dining and front door, etc. etc. *None* applies now. I really think you ought to be able to manage better. I hope I shall not find that life consists of preparing and cleaning and meals when I am home – at least at any rate not as long as you have got any maids. I shall have a lot of other work to do – more than ever before.

And in the next breath:

> Last week I went to a talk (in French) by a Madame Weill on Marie Curie. She is a French physicist – 'eminent' – who came in response to de Gaulle's appeal for scientific specialists and has been 'adopted' by Newnham and is now researching in the Cavendish – She was a pupil of Mme Curie and later researched with her in the lab. I have made several attempts to meet her, but have so far failed.

The encounter with Adrienne Weill was a turning point in Rosalind's life. Not only could she follow the lecture, in French, but she was deeply impressed by this elegant cosmopolitan woman of science and public affairs. Madame Weill became even more fascinating a few days later when she met Rosalind twice and asked her if she were related to Viscount Samuel:

> It was all rather exciting . . . Her mother, a Mrs Braunschweig (she doesn't like her name mentioned as she is still in France), is a philosopher and met Uncle Herbert frequently at conferences, etc., in Paris. She (Mme W.) has instructions to write to him about any difficulties she has in England. Her father is Braunschweig the philosopher.

Suzanne Braunschweig (1877–1946) had served as an under-secretary of state for public education in the Blum government in 1936, and had also been leader in the movement for women's suffrage. (French women did not get the vote until 1947.) Adrienne Weill had served as her mother's *chef de cabinet*. Rosalind thought an extraordinary coincidence what was no coincidence at all: 'that of all the French people now in England she should be the first I met – or that of all the families in England, she should be the most closely connected with mine. She is a delightful person, full of good stories and most interesting to talk to on any scientific or political subject . . .'

By early 1941, the Blitz intensified. On 5 January the City of London was devastated in a Sunday night sequence of attacks that left the powerful symbol of the intact dome of St Paul's Cathedral silhouetted against the blazing sky. The cathedral had been protected by volunteers – 'fire watchers' – who dealt with the incendiaries as they landed. In Cambridge Rosalind volunteered as a fire watcher. She was casting about for what she might do once she had finished her course. New possibilities were opening up. The Second World War, far more than the First, was making irreversible changes in attitudes towards women and work – half a million would be needed in industry. Even *The Times* thundered, 'The general attitude of the community to women's work required revision,' and called the barriers to women entering fields such as engineering 'intolerable and irrational'. However, Rosalind learned from a talk by 'Appointments Board Women' that her name would have to go onto the central register, which might place her in the deadly dull job of an 'experimental assistant' in the Ministry of Supply. She could only hope something better would turn up.

She was in her usual fearful mood as her final exams approached, her jitteriness exacerbated by the progress of the war. Her parents scolded her for pessimism about the war: good was bound to win over evil in the end. She denied that she was 'plunged in gloom' but refused to be cheered by the news of Rudolf Hess, Hitler's deputy party leader, landing in Scotland on 10 May 1941; it represented no change of heart, she said, merely a wish to save his skin.

Privately, Dainton, her supervisor, told Newnham that he did not expect Rosalind to get a first in her final examinations. Not for lack of ability. She had, he said, a first-class mind, and was industrious and devoted to science – if anything, too devoted – and therefore unprepared to pass on from a subject which had captured her interest to the next pressing one. Yet she was inflexible and liable to misjudge her time, answering the first questions so thoroughly that she left no time for the others.

It was an accurate forecast. Overcome with exam nerves and a bad cold, Rosalind could not revise or sleep and took what she referred to as 'dope'. She entered the examination slowed and dulled, and did indeed misjudge her time. In two papers she completed only two questions out of a required three, finding herself 'almost incapable of thinking'. Her sense of ineptitude was particularly annoying because, once again, the papers were 'almost too good to be true'.

She blamed the head cold and the sedative: 'I'm sure it sounds silly to say so after the event when I did not think so before, but I really feel certain that I could have got a First on those papers if I had been fit. Anyway there is now absolutely no question of a First.' Out of the question too, she believed, was any chance of a government grant for research. She did not expect the college would sympathise with her illness as there was great competition for the single available studentship – 'not sleeping will naturally be considered my fault, the result of trying to do too much'.

Once more, however, she had not done as badly as she feared, nor was she blamed for her insomnia. Sleep problems are common before exams. It was known that the great William Bragg could not sleep the night before his tripos. But she had not done as well as she ought to have done. Her degree was a good second – a creditable conclusion to her three years – and she was told privately that in the physical chemistry exam she had come out on top. Although she had failed to achieve a 'double first', a first in both parts of the tripos, her overall performance was distinctive enough for Newnham, with Dainton's endorsement of her qualities of intellect and tenacity, to award her a college scholarship of £15, to remain for a further year, and for the Department of

Scientific and Industrial Research to give her the research grant she hoped for.

Cambridge, in spite of the war, did almost everything for Rosalind that a good university should. It changed her life. It gave her a profession and a personal philosophy. It enabled her to distance herself from her parents and become a mature adult with a sharp political and social conscience. That she achieved all this amidst the self-doubt, the confrontations to which she was prone, and the terrors of war in the years when Britain stood alone is a measure of her inner steel. What three years at Cambridge did not do was end her astonishing ignorance about sex. Rosalind confessed to her cousin Irene Franklin (Ursula's sister) who had just become engaged and came to visit her that summer, that she had never been kissed. The talk turned to having babies. The cousins discovered that although each knew vaguely how a baby was born, neither knew how the ovum was fertilised. (A few months later, Irene informed Rosalind that her fiancé had enlightened her. Rosalind was wiser too. She said she had asked a medical student.) But such things were remote from her mind. She was now ready to be a working scientist. The question was where to work.

FIVE

Holes in Coal

(1941–46)

R.G.W. NORRISH, FRS, Professor of Physical Chemistry, holder of the Royal Society's distinguished Davy medal, was having a bad war. In his forties, recognised as one of the pioneers of photochemistry, much later to win a Nobel prize, he was under great strain. His laboratory had shrunk in size, with so many men away, and had no clear objectives. His wife had taken their twin daughters to Devon to escape possible bombing. He was left to fend for himself in Emmanuel College, drinking heavily. (The college fellows had wisely stocked up the cellars to withstand the siege.) This failing cost him a high security rating; he was not entrusted with serious war work. His discontents found outlet in what Fred Dainton, then senior research adviser to the Physical Chemistry Laboratory, observed as 'bad tempered and autocratic treatment of juniors', of whom Rosalind had the misfortune to become one.

Norrish gave her what Dainton, having supervised her the previous year, could see was a trivial problem – the polymerisation of formic acid and acetaldehyde. Rosalind would have preferred something that more directly related to defeating Hitler. In the room next to hers, salvaged fuel from crashed German planes was being analysed with spectroscopic techniques.

Her St Paul's friends were clearly aiding the war effort: Anne Crawford at the Bristol Aircraft Corporation, Jean Kerslake at the code-breaking centre at Bletchley Park. Her cousin Ursula

Franklin was in the women's army service – the ATS. However, when Newnham awarded Rosalind a fourth-year scholarship, she was spared military service and allowed to remain at university, to her father's dismay. Yet what exactly she ought to have been doing instead was hard for him to say, as a woman's place in the war effort had not been defined.

At times she must have felt that no one was nice to her. Norrish gave her a small dark room to work in even though (as Dainton knew) she suffered from claustrophobia. When she asked her college, where she was still living in college and serving as an air raid warden, if she could come back a few days early in the autumn, her old adversary, Mrs Palmer, was disagreeable. First she declared, as if speaking to a schoolgirl, that Rosalind was old enough to make arrangements for herself until term began, then allotted her a horrid room – 'on the pretext', Rosalind wrote home, 'that I managed the blackout badly where I had lots of windows'.

The solution was a rented room on Mill Road, a working-class area near the railway station: 45 shillings a week, heating excluded. Living alone for the first time, she proclaimed the virtues of solitude: 'I've never had so much time for reading. I read at and after every meal, and in the evenings when I'm doing nothing else.'

She had no difficulty filling her time. She finished a dress that she had begun making a year and a half before. She made meals in her room and invited friends in, treating one to a lecture on the nutritional value and cheapness of sprats (small fish). She was trying to get by on her scholarship money, and living on a shoestring suited her.

She also listened to the radio. For her twenty-first birthday, her aunt, Mamie Bentwich, had given her at her request, 'a baby wireless . . . the sort that goes off the light'. Up to 50 per cent of the total population listened to the BBC nine o'clock evening news. As winter approached, the Germans were driving towards Moscow and great hope was placed in the RAF's night raids over Germany. Only the men of Bomber Command knew how inaccurate these were; in the early years of the war, only one bomb in three fell within five miles of the target. One pilot told the Cambridge crystallographer, J.D. Bernal, then serving as Lord Louis Mountbatten's scientific adviser: 'You can think it damned lucky, old boy, that we drop the bombs in the right country!'

In her evening reading, Rosalind tackled the just-published memoir *Wanderer Between Two Worlds* by her Aunt Mamie's husband, Norman Bentwich, but did not like it. 'Considering the experiences and opportunities he's had, it could have been more interesting,' she told her parents, 'and I think the catalogue of leading Jews, nearly all titled, whom he's met, are most objectionable.' Impatience with self-indulgent prose prevented her from finishing Virginia Woolf's *To the Lighthouse*: 'I like long sentences when well put together, but hers are so arranged that the beginning is meaningless until the end is reached, which I consider quite unjustifiable.'

When her work was going badly and tension was in the air, however, Rosalind found herself once again 'quite incapable' of concentrating as when, early on, she spotted a fundamental error in the project Norrish had given her. Dainton, who had recommended her for the research scholarship and had a high opinion of her ability, agreed that she was right. The long-brewing crisis came to a head early in 1942. Rosalind wrote up a summary of her findings showing why it was impossible to get the result Norrish expected. The professor, however, declined to read what she had written, refused to change his approach and ordered her to repeat the experiments. There was no alternative but a showdown:

> When I stood up to him he became most offensive and we had a first-class row – in fact, several. I have had to give in for the present but I think it is a good thing to have stood up to him for a time and he has made me despise him so completely that I shall be quite impervious to anything he may say to me in the future. He simply gave me an immense feeling of superiority in his presence.

Confrontation when cornered was Rosalind's tactic. The alternative – passive acquiescence in something she knew to be wrong – was intolerable, totally contradictory to her faith in the provable truth of science.

She had been trained in a hard school. A civil servant who was working with Ellis Franklin on refugee applications at the Home Office witnessed the training in action. He was spending a weekend with the Franklins at their temporary home in Radlett, when Rosalind suddenly turned up from Cambridge. She and her father immediately engaged in fierce political debate – 'he on the right, she on the left' – in a manner so heated that the visitor, K.C. Paice, thought that Rosalind must be an 'uncompromising Communist', which seemed 'a violent contrast with the extreme luxury in which she appeared to have been raised'. He observed that neither Rosalind nor her father tolerated dissent from their views: Ellis was a 'domestic martinet'; she was 'her father's daughter'. However, the visitor also noted that Rosalind was 'strikingly good-looking'.

Many people over the years observed similarities between Rosalind and her father. The same description (in essence, 'did not suffer fools gladly') was repeatedly used for one or the other. Yet Ellis Franklin seems to have paid no professional price for combativeness. The same was not true for his daughter.

Rosalind's stand did, nonetheless, wring a new project out of Norrish – 'not thrilling, but it MUST be better than the last'. She also made new friends, thanks to the French scientist she so admired, Adrienne Weill, who had gathered a group of French refugees to live at a hostel near the Cam. Spending time with them, she discovered an unexpected dimension to her personality.

> I don't know whether I meet here a particularly select French crowd but I always revel in their company. Their standard of everyday conversation is vastly superior to that of any English gathering I have been in and they are all so much more quick-witted and alive – I love listening to their

> language . . . though I find myself unable to take part, the pace is much too fast for me . . . I'm thinking very seriously of moving to 12 Mill Lane – Mme Weill's place.

In the summer of 1942, she had to decide whether to apply for permission to stay on at Cambridge as a research student in a programme that would lead to a PhD or risk being drafted into the Ministry of Supply or some other government agency.

Ellis Franklin did not like the sound of it: his daughter strolling the groves of academe while his elder sons were putting their lives at risk for their country. Her stay in Cambridge seemed to him like his German year in Breslau. Taking the risk of arousing her fierce temper, he wrote Rosalind what he thought, adding several other complaints on his mind and enclosing the gift of one pound to soften the blow. Return fire was not long in coming. 'Dear Father,' she began, on 1 June 1942, 'I certainly don't resent criticism as such.' But:

> On one point you are quite unjust – I don't know where you got the idea that I'd 'complained' about giving up a PhD for war work. When I first applied to do research here a year ago, I was asked whether I wanted war work, and said I did. I was led to believe that the first problem I had was war work. I soon found that I had been deceived, and since then have made repeated requests to Norrish for war work – it's one of the many things over which we have differed – and I have explicitly stated on several occasions against the advice of my elders and betters, that I would rather have war work now and a PhD later.

Another way in which she had offended him was by saying that her mother's life was dull. She was sorry but rather than apologising, stood her ground: 'I am extremely sorry that it should be so, but, as you say, she is tied to the house most of the time, doing all sorts of domestic work alone, and I can't see how it can be anything but dull . . .'

She herself was bored with her work, she said. She despised her professor and disliked her colleagues who 'resent and generally ignore my presence'. Therefore, it should not be surprising that she was, as he accused her, thrown back upon herself. In any event, her life at present bore little resemblance to his at Breslau.

Diversion in a difficult time was once more supplied by members of her family. Twelve-year-old Jenifer came for a visit and Rosalind put herself out to give her sister a good time – indeed, providing one of the most memorable weekends of Jenifer's childhood. Rosalind showed her sister the grand buildings of Cambridge as well as her own laboratory where she demonstrated blowing and assembling glassware. They then went to see the wartime epic, *49th Parallel,* starring Laurence Olivier and Leslie Howard. On another occasion her aunt, Mamie Bentwich, accepted being put up in 'a room full of junk and two beds kept for casual visitors'. In the evening they went to the theatre and at breakfast sat in their dressing gowns and gossiped. Mamie reassured Rosalind that the war, in spite of the German advance on Moscow, would not last for another ten years.

At the end of the summer Rosalind packed up her wireless and shifted herself away from the bleak room near the station and down to 12 Mill Lane, to join Adrienne Weill's small hostel. She hesitated slightly before moving. She knew enough about herself to know that she liked people better when she didn't have to live with them. However, she felt that she had become a more tolerant person, although she knew, she said, her parents would not agree.

Adrienne Weill remained impressive. The French scientist was now working at the Cavendish Laboratory on a salary from the Ministry of Supply, thanks to the ingenuity of the Cavendish's director, Lawrence Bragg, who found her a subject that could be considered war work but not too secret to be dealt with by a foreigner. For Rosalind the chance to speak French on a daily basis was hard to resist, and she soon decided she had found the ideal existence. She got along well with the other members of the hostel. Marianne Weill, Adrienne's daughter, studying at the Perse School in Cambridge, formed a teenager's view of Rosalind: 'she was extremely kind, good and serious; you didn't see her smile very often'.

The move to Mill Lane coincided with Rosalind's twenty-second birthday and a parental deposit of five pounds in the bank, for which she was grateful: 'as I can never see it so shan't know when it's gone, and will therefore feel justified in spending anything on everything for ever'.

So should she stay in Cambridge? The choice was between applying for permission to stay on, continue research and work towards her doctorate. Or to see what jobs might be offered her in the wartime lottery: the Ministry of Fuel or Ministry of Supply, or even women's military service. The decision was forced by an announcement from the Ministry of Labour that all women research students, even those doing war-related work, were to be 'de-reserved' – that is, made eligible to be called into military service. Rosalind, who never complained about inequality of treatment of the sexes, was surprised that all male researchers of comparable status were to be allowed to keep on with their university work. Norrish astonished her by urging her to apply to remain 'reserved' and thus able to stay on. Recognising (although he never told her directly) that she was a brilliant experimentalist, he painted a bleak picture of the horrors of industry. 'I could hardly keep from smiling,' she told her mother. Working with a commercial company might be tolerable, but not if the war dragged on for years:

> In industry there undoubtedly *are* better jobs but they never go to women. If it's only a matter of a year or so it doesn't much matter, but if it's 5 or 10 – it probably would be better here. I favour Cambridge . . . if the job offered is reasonable. As long as one stays in a university – even on

> a utilitarian work – it is science for knowledge – I'm so afraid that in industry, I should find only science for money.

It was not easy to find the right environment in which to continue physical chemistry. Rosalind, struggling with all scientists' problem of conveying the substance of their work to family and friends, tried to explain to her parents that her particular line of research needed probably more apparatus than any other branch of chemistry. She had for her exclusive use about £100 worth of equipment, many more thousands of pounds' worth at her disposal – 'and a daily supply of liquid air is essential!'

The situation resolved itself in the form of a chance to do research in a government laboratory in the borough of Kingston-upon-Thames, on the southwestern edge of London. To Rosalind, a Londoner, the suburbs were as unappealing as the provinces. To her parents, who had now reoccupied their old home in Bayswater, she said she would take the job if it were offered.

> I think the work sounds not too bad, but I don't like the idea of the place – miles from anything, without even the consolation of a lunch-hour in town – it'll be lunch in the lab, with lab people, all horribly shut off. But the alternatives are probably worse. I did not seriously consider the Ministry of Fuel job. I'm sure I should be very much more efficient in administrative work, but I should lose touch and never be able to go back to lab work, which would be much more exciting if only one could succeed.

She felt she had failed twice in Cambridge – in not getting a first-class degree and not achieving anything in her work with Norrish. When in August 1942 she accepted the job in Kingston, Dainton, who now knew her better, thought she was absolutely right to get out. Married that year, he and his wife, a zoologist from Newnham, were touched by the care and skill Rosalind took in choosing a wedding present she knew they would like. He thought it a tragedy that she had had to work for Norrish, with whom nobody could get along at that period in his life. In any case, she now had something that met all requirements: her own, her father's and her country's.

If she had had an ideal existence before, she improved on it now: a large house on Putney Common, shared with her cousin Irene Franklin and a friend, with a small private garden, a daily woman to help with the cleaning, and the rest left to themselves. As Rosalind had the shortest working hours, she did most of the housekeeping and cooking. Despite what she had said about her mother's life, she thoroughly enjoyed housewifery. She wrote to their young former Austrian lodger, Evi Eisenstadter Ellis, now living in Chicago (where the family, changing their name, chose 'Ellis' as a surname out of respect for Ellis Franklin who had saved their lives). 'Even washing-up ceases to be unpleasant,' she said, 'when you do it in your own place and at your own convenience.'

Peace of mind came from a satisfying job at last. The new British Coal Utilisation Research Association (BCURA) under its first director, Dr D.H. Bangham, had assembled a staff of graduate physicists direct from universities to study coal and charcoal. At BCURA (pronounced 'B'Cura' by those who worked there) young researchers were allowed to do original work in a way that would not have been possible before the war. (Charcoal, used in gas masks during the First World War, had saved thousands of lives and was a component in the masks that Rosalind and other students carried in Cambridge.)

Most British coal is derived from fern-like carboniferous plants which are transformed by degrees towards almost inorganic coal by squeezing – the process of 'coalification'. The question Rosalind now addressed was why some kinds of coal are much more impervious to penetration by gas or water than others. She worked on bituminous and anthracite coals from Kent, north-east England, South Wales and Ireland. Using helium to see how much could pass through the imperceptible apertures in the various cellular structures, she tested the change in porosity under temperatures as high as 1,000° Celsius. She had apparatus galore – lamps, rollers, furnaces and a good supply of dry nitrogen – to conduct her experiments. As she lowered the temperature and raised it again, measuring the shrinkage in coals and carbons, she developed theories that would make her international reputation.

With the title of Assistant Research Officer, she had a fairly authoritative way about her. One day when she went into the machine shop – a vital spot in any laboratory reliant on big apparatus – and found signs declaring it out of bounds to non-certified personnel, she simply turned the signs around, and used the equipment anyway.

At home in Putney, Rosalind and Irene volunteered as air raid wardens. Their duties consisted of inspecting the blackout – wearing tin hats and cycling or walking around their assigned patch, looking for violations – or sitting in the 'Post' for two-hour stints. Rosalind was fearless, up to a point. She was bold in venturing alone across the dark open common during an air raid, but glad that it was Irene and not she who had to go into a bombed-out house to rescue people trapped in the cellar. Irene blamed her cousin's claustrophobia yet admired Rosalind's courage in walking across the common, as she herself was too scared to do that. In many ways, from Irene's point of view, Rosalind was 'too good at everything: work, sport, looks, cooking'.

Strenuous, even dangerous, holidays were Rosalind's relaxation. In the summer of 1943 or 1944, with her old friend Anne Crawford, she went climbing in Snowdonia, North Wales. The predictable crisis arrived; the mist came down as they were halfway up the ridge of Crib Goch, with steep drops on either side. Anne inched her way to safety – 'driven more by my fear of Rosalind's tongue than of falling over the edge'. Rosalind's ability to retain close friends in spite of her capacity for scorn was remarkable.

On another holiday in North Wales, this time with Jean Kerslake, they tackled the highest peaks in the same spirit in which they had paced each other at St Paul's Girls' School. One day the pair set off on a ten-mile walk to the shores of Llyn Dinas. When they arrived, they encountered two young foresters who were living in a caravan nearby. The men gave them a lift and announced themselves as Welsh nationalists. 'There are only two of us,' they said, 'but it is a beginning.' In the hot sun, the lake looked inviting. None of them had a bathing suit so they stripped off and swam naked. Rosalind did not hesitate. Her attitude, to Jean, was 'She wanted a swim, so she had one.'

The quartet repeated the escapade the next day, cycling down to the coast at Criccieth where they swam, again without burden of suits. Even so, when Jean and one of the young men took a fancy to each other and hung behind, they returned to the caravan to find an annoyed and puzzled Rosalind, waiting with the other young man. She seemed not to comprehend why the two had wanted to be alone.

The Putney ménage broke up at the end of 1943. Rosalind had been living at home for a time because of an attack of jaundice; Irene had married and was expecting a baby (or in Ellis Franklin's arch words, 'Irene will be taking on the duties of motherhood in the spring'). He looked forward to having Rosalind remain under his roof, and told his son Colin, serving with the Royal Navy in the Far East, 'the house will be that much the brighter'.

Commuting out to Kingston and back every day meant a long slow journey but she was glad of the shelter and solidity of the family home during the new round of air raids. The first of the V-1s – pilotless flying bombs – struck on 12 June 1944, and just after D-Day, and on 8 September, the first V-2s. 'Of course for anyone who suffers directly, any type of bomb is equally bad,' Rosalind commented to Evi, safe in the United States. 'But I think most people agree that for those who are not hit, the present type of raid is much less worrying than the older variety.' Her young sister Jenifer moved out to Chartridge, Irene and her new baby in the flat over the garage. Rosalind joined them at weekends and was surprised to see her young sister's intuitive gift for holding the infant and calming its crying.

Rosalind was, and remained, a curious blend of compliance and self-reliance. Her parents did not dream of trying to choose her friends or to stop her travelling freely on hazardous journeys. On the other hand, they expected her when at home to perform her filial duties, and she did, even when these included sitting impatiently with them through a very long carol concert at the Albert Hall.

By the spring of 1945 ending the war was just a matter of time. Adrienne Weill had returned to Paris after the Liberation in 1944, there. Her daughter who remained for a time in London was entertained at the Franklins' and quite impressed by the way they lived, especially by the British breakfast spread on the sideboard. In January 1945 Ellis Franklin received the Order of the British Empire for his work at the Home Office Ministry of Home Security, and resumed full-time work at Keyser's bank and at the Working Men's College. With some help from his brother-in-law Norman Bentwich, he set about reorganising the St Petersburgh Place synagogue, hoping to convince younger Jews that they could be Jewish as well as English, but also to counteract the activities of 'alien Jews who have captured some of the organisations here, and are running them for purely Zionist motives'. He was extremely upset all the while at the general scepticism in London about reports emerging about massacres of the Jews in eastern Europe. (As Rosalind was living at home, there is no correspondence to reflect her own opinions but one assumes that on the British denial of the existence of the extermination camps, she agreed with her father.)

Rosalind, who disliked abandon in any form, disapproved of the mad and, as she saw it, premature euphoria of VE Day in May 1945. Two months later she did, however, anticipate the war's end by going out to buy a pair of shoes at Lilley & Skinner. Finding a queue of 200 people, and long lines at two other shoe shops, she went to Harrods instead and bought some excellent chocolate peppermint creams.

She had begun to look for a post-war job that would take her further afield than BCURA. More immediately, she had to finish her thesis on coal. Briefly she moved back to Cambridge, staying in her old college. She savoured the good weather, the freedom, the beauty of the gardens, and was tempted to return to take up the grant she had been offered to work with Norrish.

Norrish had forgiven her, but not she him. Her parents, trying to advise her on getting along with people better, ventured that such relationships should be impersonal. It was not a *job* he was offering her, Rosalind retorted: 'He's merely expressed willingness to have me work for him for a year as an unpaid stooge.' She would have to do whatever he told her to do 'and he is stupid, bigoted, deceitful, ill-mannered and tyrannical'. Thus, having talked herself out of returning to Cambridge, she sent her thesis – 'The physical chemistry of solid organic colloids with special reference to coal and related materials' – off to the typist.

On 26 July 1945, a general election brought the Labour Party, under Clement Attlee, to power, turning out Winston Churchill and the Conservatives with a resounding majority of 146. (In the new Parliament Viscount Samuel became Liberal Party leader in the House of Lords.) In early August Rosalind and Jenifer went to visit Nannie Griffiths at Church Stretton in Shropshire. There when the news came of the atomic bomb dropped on Hiroshima, Nannie and Jenifer listened while Rosalind explained what it meant.

The post-war world taking shape in 1945 saw the admission of the first women to the Royal Society, for nearly three centuries the citadel of Britain's scientific elite. There were just two: Kathleen Lonsdale, a crystallographer with the Royal Institution in London, and Marjory Stephenson from Cambridge, a pioneer in the field of chemical microbiology. Forty-three years had passed since the Society threw out the nomination of the first to be proposed, Hertha Ayrton, an engineer and physicist, on the ground that as a married woman, she was not a legal person and therefore could not be a Fellow of a body governed by statute.

In that light, 1945 might also be described as the year in which Lise Meitner did not get the Nobel prize. A Jewish refugee from Berlin living in Stockholm, she was arguably the most distinguished woman scientist of the time. In 1939 with her nephew Otto Frisch, she had set out the concept of nuclear fission, an outgrowth of her three decades of work on radioactivity done with Otto Hahn in Berlin. In 1945, the rumours preceding the annual Nobel announcement, had it that the physics prize would go to Meitner. It did not. Hahn won the chemistry prize, but there was nothing for Meitner that year, nor in any of the subsequent years when her name was put forth. Some press reports described her as Hahn's subordinate.

The goal reached at last – PhD Cantab., 1945 in physical chemistry – and her first scientific paper on its way to publication, but then what? The possibilities ranged from Aberdeen to Abyssinia (there was an opening for an assistant mistress to teach physical training) as Rosalind searched for the right next step. Turning to Adrienne Weill, she wrote one of her forceful, humorous letters, and summed up her hope:

> If ever you hear of anybody anxious for the services of a physical chemist who knows very little about physical chemistry but a lot about the holes in coal, please let me know.

She did know quite a lot about holes in coal. Her first paper, titled 'Thermal Expansion of Coals and Carbonised Coals,' was written with her BCURA boss, D.H. Bangham and appeared in the British journal, *Transactions of the Faraday Society*, in 1946. At BCURA, she had developed the hypothesis of 'molecular sieves' – that is, that various types of coal have porous properties to a greater or lesser degree. By performing careful mathematical calculations on the porosity and carbon content of her specimen coals – also by heating, wetting, grinding, measuring and examining them under the electron microscope – she concluded that the change in physical properties comes from their gradual squeezing together. The understanding of such hard substances offered intriguing industrial applications.

In the summer of 1946, with foreign travel possible once more, Rosalind packed her Norwegian hiking boots and, with Jean Kerslake, set off for the French Alps, stopping at Adrienne's flat in Paris on the way. Her usual careful advance planning had elicited a warning from the manager of a hostel in the Hautes Alpes: 'I make you notice that, in the case you would not know it, the French youth hostels are generally too uncomfortable. They can only fit very sportive persons who are free of any prejudice.'

Rosalind Franklin, in other words. The lack of indoor plumbing and the hard floor were positive attractions, as were the other hardships they encountered on the way – shabby hotels, the rickety bridges hastily thrown up to replace those destroyed during the war, and a stomach bug. The young women took the advice of a guide, who taught them the use of ice axes and ropes, and climbed to the summit of the steep Aiguille Pers. It took them so much longer than they had planned that when they returned, they found that the modest hotel where they were then staying (and which had been a little surprised to find two English girls travelling alone) was organising a search party. Embarrassed but forgiven, Rosalind and Jean drank wine with the crowd and passed round the cigarettes they had brought from England. On returning home, Rosalind wrote to her mother, 'I am quite sure I could wander happily in France for ever. I love the people, the country and the food.'

Did she ever fall in love with an individual? Jean Kerslake recalled that as much time as she spent with Rosalind, that summer and over many years, they never spoke about romance or sex, nor gossip about the love life of any of their friends. 'She did not talk about men as the rest of us did,' said Jean, looking back, 'and it seemed impossible to break down her reserve.'

The possibility naturally occurred to Rosalind's friends and some of her family that she might have been inclined towards the love of women. They all ruled it out because of too much evidence to the contrary. In London in the autumn of 1946 Jean, with another school friend Celia Martin, joined Rosalind when she showed the tourist sights of London to two French scientists who had come to London for a Royal Institution conference on carbon. To them, it was obvious, seeing Rosalind animated and chatting happily away in French, that she was strongly attracted to the younger of the two, Jacques Mering. Seeing her unfamiliar sparkle, her friends thought it was most unfortunate that, as they gathered, Mering already had a wife and a mistress. They would have liked to see Rosalind settle down as they and their other St Paul's friends were doing. In Jean's opinion, 'I think she would have liked to have a relationship with a man, but had not the remotest idea of how to cope with them or where to start.'

What Rosalind did know how to do, at the age of twenty-six, was to speak forcefully in public. At the Royal Institution's meeting she presented a paper, and then rose to her feet to point out the errors in someone's measurements of X-ray powder diagrams. She was self-confident and forthright, or, in the words of the head of crystallography at Birkbeck College, Harry Carlisle, 'abrupt and peremptory'. There was no answer to her comments, Carlisle granted. However, he observed (after getting to know her much better), 'her characteristic of being forthright when she knew she was on firm ground sometimes gained her enemies'.

But not among the two French visitors. Mering and his colleague, Marcel Mathieu, were crystallographers from a French government laboratory; Mathieu was a close friend of Adrienne Weill's, and at her suggestion had looked up Rosalind when they came to London.

Within weeks Rosalind had the offer of her dreams: a

challenging job requiring the services of a physical chemist to study holes in coal in Paris. Adrienne Weill had turned out to be the fairy godmother she had seemed.

SIX

Woman of the Left Bank

(1947–49)

THE LABORATOIRE CENTRAL DES SERVICES CHIMIQUES DE L'ETAT, at 12 Quai Henri IV in the fourth arrondissement, downriver from Notre Dame, was the right place for a francophile with a taste for the analysis of awkward crystals. It was a government laboratory, originally under the French Ordnance Ministry (Ministère de Poudre) but with a post-war research programme aimed at industrial applications. In the 'labo', as the staff called it, there were fifteen *chercheurs*, of whom Rosalind was one, and half a dozen technicians, under the direction of Jacques Mering. They were all terribly impressed by Rosalind, Mering not least. He immediately saw that she knew what she was doing and that she was very good at delicate experimental work. His speciality was the use of X-rays to study the internal structure of crystals known as 'disordered' because of imperfections in the arrangement of their molecules. Rosalind, in her studies of the carbon structure at BCURA, had used physical-chemical techniques such as heating and grinding to measure the porosity of coal and carbon. Now Mering could teach her how to employ X-ray diffraction to look at the internal organisation of charcoal and clay.

Mering (pronounced Mer-eeng – no nasal) had been trained by Marcel Mathieu, the good friend of Adrienne Weill who had been responsible for bringing Rosalind to Paris. In 1946 Mathieu, a witty, expansive, brilliant, card-carrying Communist, was working in the department of materials under the Ministry of

Defence, and kept a paternal eye on Adrienne's protégée. He had been trained in crystallography at the Royal Institution in London in the 1920s, under the elder (William Henry) Bragg.

In applying X-ray analysis to amorphous substances, Rosalind had, at last, the satisfaction of building on what she had been doing before. Mering used highly monochromatic (single wavelength) and finely focused X-rays to take low-angle photographs, revealing bands of varying intensity rather than the sharp spots made by more ordered crystals. Disordered matter had become a French speciality, analysed by techniques then not widely used in other countries. With her skill at chemical preparation, Rosalind was soon on her way to detecting and clarifying the fundamental difference between the carbons that turned into graphite on heating and those that did not.

Mering was 'Monsieur', not 'Professeur'. A Russian-born Jew, he had come to France early in life and was working at the Labo Central when the Nazis occupied Paris. The Ordnance Ministry, in a move to protect basic research pursuits as well as the distinguished French-Jewish scientists on its staff, shifted work out of Paris. Mering thus moved to the University of Grenoble, where he set up an X-ray laboratory. At great risk to himself, Mering did not declare his Jewishness, a fact which was supposed to be stamped in his identity card, and thus lived without proper identity papers throughout the war. When he came back to Paris, without seeking further academic qualifications, he resumed his research at the Laboratoire Central. He was just one of many returning refugees, for whom the first priority was, in the words of Marianne Weill, Adrienne's daughter, 'knowing who was alive and who was dead'. Marianne, who had returned from England with her mother, was now at the Sorbonne, while Adrienne was a metallurgist at the French government laboratory for naval research.

When she arrived in Paris in February 1947, Rosalind quickly became, in D.H. Lawrence's word, 'unEnglished'. Apart from her

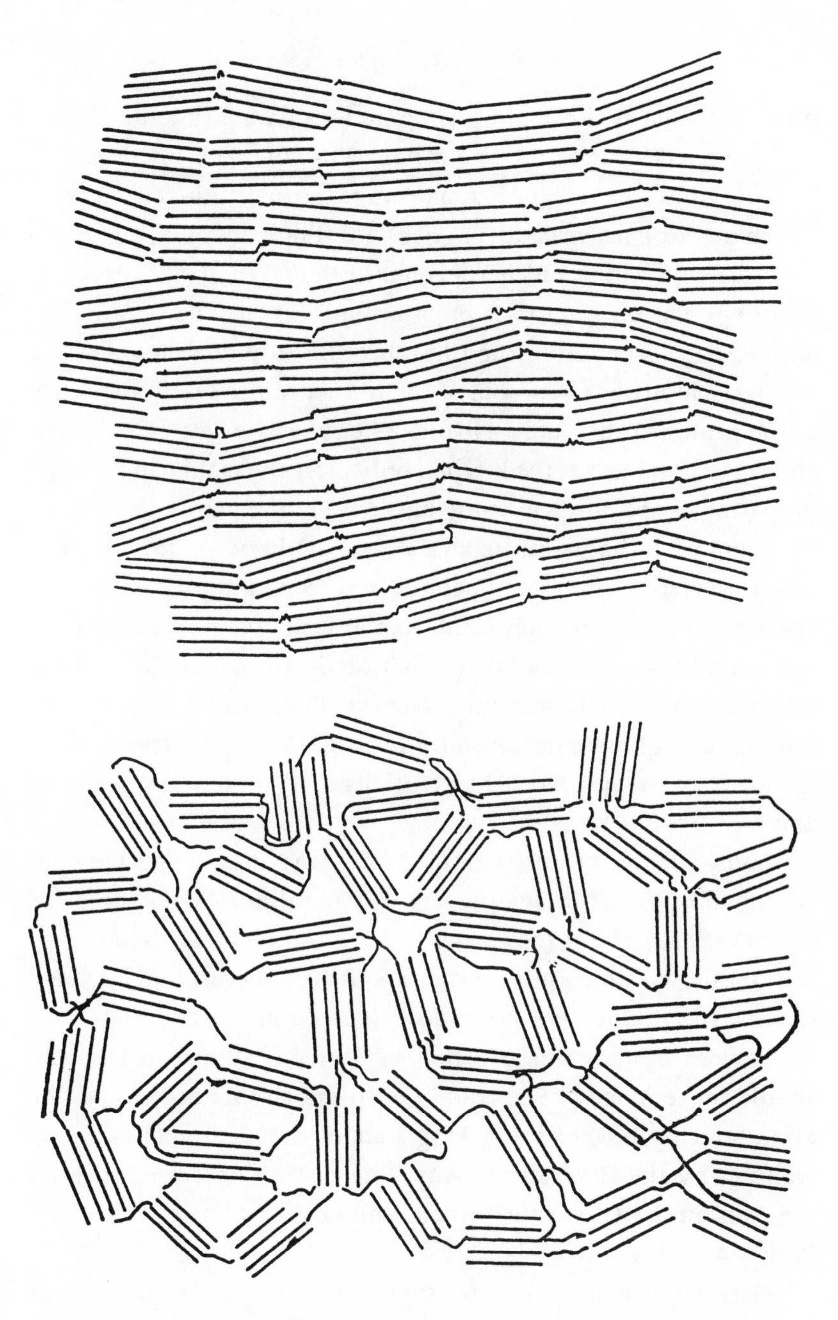

In Paris Rosalind found, by heating carbons of different origin to as high as 3000 degrees C that graphitising carbons (*top*) and non-graphitising carbons (*bottom*) formed two distinct classes. In this schematic representation, the non-graphitising carbons are distinguished by a rigid, finely porous mass. This discovery had important industrial applications.

breakfast, that is. Resuming her practice of weekly letters home with detailed information about her well-being and finances, she recited what she ate every morning: bread (rationed), butter (black market), marmalade (English), tea (English), milk, and fruit.

Adrienne had found her a room in the sixth arrondissement, in a huge flat on the top floor of a house in the rue Garancière, around the corner from the Church of St Sulpice. The owner, a widow, had made a bedroom of the library stuffed with the relics of the professor, her late husband. The rules were strict. No noise after 9.30 p.m.; use of the kitchen only after the maid had finished preparing the widow's evening meal; use of the bathroom (that is, the room with the bathtub) once a week. Otherwise, the washing facilities were a tin basin behind a screen. But the room was spacious and the location ideal, in the heart of the Left Bank, a few picturesque streets away from St Germain des Prés where tourists hovered around the Cafés de Flore and Deux Magots hoping for a glimpse of the pair whom the *New York Times* called 'France's No 1 and No 2 Existentialists' – Simone de Beauvoir and Jean-Paul Sartre

Considering its restrictions, Adrienne had expected the accommodation to be temporary, but Rosalind stayed there for the next three of her four years in Paris. She felt secure, and the rent was low – the equivalent of about £3 a month, only a third of what she would have to pay elsewhere because, she recognised, 'the owner is more anxious that I be respectable than that she should make money'. She could get to the lab in about a quarter of an hour by bus, or by making a slight detour in half an hour, walking by the river all the way. She cast a researcher's eye on the constant haze over the Seine: 'while London mists are yellow, Paris's are blue'.

Her parents were not so sure about the respectability of Rosalind's new arrangements. Was she not lowering her standard of living? How could she live on her salary? And how important was her work? She fell easily back into her old epistolary mode:

> Of course my standard of living is lower than at home . . . Of course I appreciate conventional comforts and of course I would rather the food situation here were normal . . . but provided one does not go below certain minimum none of these things are of supreme importance to me . . . I find life interesting . . . I have good friends though my circle is naturally smaller than in London but I find infinite kind ness and goodwill among the people I work with. All this is far more important than a large meat ration or more frequent baths.

Her job was in a government research establishment, with no immediate industrial objective. Some people might call it 'pure research', while others could argue 'that there is no such thing as "pure research" since all scientific advance is ultimately useful'. She was paid according to the salary scale for French government workers, with the freedom and facilities 'to work on my own ideas – and anybody else's I may be able to borrow'. Her earnings of approximately £5 a week were sufficient for her ordinary living expenses. Extras such as holidays and clothes had to come out of what money she brought with her. Besides: 'One only feels rich or poor in relation to the people one mixes with, and as all my friends are in the same circle . . . or worse off because they have not resources in England.'

Resources in England could not help very much. In 1945 John Maynard Keynes wrung a $4 billion loan from the US Treasury instead of the $5 billion outright gift he had expected in recompense for Britain's brave and successful resistance to Hitler. The loan was supposed to last four years but was gone in less than two. In consequence, the newly nationalised Bank of England slapped stringent restrictions on taking sterling abroad. Any gifts and even personal possessions were subject to rigorous examination by Customs officials on both sides of the Channel. Even Rosalind's sewing machine was held up for two months.

For Rosalind, such restrictions merely added to the zest of her new existence. Not only was her work fascinating but the crowd at the lab were fun. Every day they crossed the river to lunch at a small restaurant, Chez Solange, overlooking the Ecole de Physique et Chimie where Pierre and Marie Curie had discovered radium. Many of the lunchtime regulars had been in the Resistance, many were communists and all were glad to be part of *'la bande de Solange'* or *'les gens de Chez Solange'*.

After lunch the band would move into the Physics and Chemistry School for a ritual known as *'les Cafés de PC'* – coffee brewed in a laboratory flask and served in evaporating dishes while the animated conversation rolled on – in a spirit described by one of their number as 'liberal, Cartesian laced with atheism and defence of the rights of man': in other words, French intellectuals playing the part of French intellectuals. There was plenty to discuss and the women engaged as equals, with no fear of condescension. Paris was at the height of its post-war political ferment, looking for a 'third way' between the Soviet Union and the United States. To the band of *camarades* General de Gaulle seemed a right-wing threat. They also talked science. Rosalind would show her X-ray photographs to Mering, who would interpret them. On occasion, English crystallographers would drop in, notably J.D. Bernal, a good friend of Marcel Mathieu's.

In that milieu Rosalind seemed 'a sound and cheerful young lady' who spoke perfect French with a very slight English accent and who was totally committed to her work. On occasion the group would go for a swim at a pool nearby. At night they sometimes went dancing, Rosalind riding pillion on someone's motor cycle. UnEnglished, she was having the time of her life.

She alerted her parents to a gallic indulgence of temperament:

> I've been told several times my French is best when I'm angry. I had a glorious row this a.m. in a shoe shop that sold me a pair of straps of wildly different length. I enjoyed it immensely. I haven't felt the language flow so freely since the row which Roly witnessed with the hotel keeper in Pralognan. It's odd but I couldn't do it in English with a straight face. Here a battle of words is a sort of game which

> nobody takes seriously. I hate it when it happens to people I know but with disagreeable hotel keepers and shopkeepers it can be a satisfying pastime.

Then in the next breath:

> Are skirts seriously getting longer in London? Here any thing that hasn't been lengthened is really coming to look ridiculous – a most tiresome business. And new things are only 8–10 inches from the ground (not mine). People take it seriously (and that's most people) chop up their skirts and put extra bits in and cut down their waists to ¾ length.

Paris affected Rosalind as it does many women who suddenly find themselves trying harder to match the scarf to the jacket. She had always taken clothes seriously and combined good taste with restraint and propriety but now she was challenged to change. The imperative to smarten up became urgent when – world news in 1947 – Christian Dior introduced his 'New Look'. Utterly out of step with bankrupt Britain and the masculine look of women in uniform, it set off an international controversy with its exaggerated femininity – nipped-in waistlines, small shoulders and extravagant use of material in long voluminous skirts. Rosalind was an immediate convert. 'I've had dress material made into a handsome [pronounced] NEV Look frock and bought some nylons,' she wrote home, 'but need a new petticoat forty-four inches long to wear under this long-line.' Could her mother's seamstress make her one out of parachute silk (more easily available in London than in Paris) and also some silk underclothes so she could throw away her pre-war collection? She sent, along with a diagram of the shape of knickers she wanted, her measurements: bust 34, waist 27, hips 38. She also followed prevailing post-war style by registering her small waist with a neat belt.

A lab visit to Nancy with Mering at her elbow, shows her in a classic New Look bold-checked suit: with small neat shoulders, a dramatically cinched waist, and exaggerated full skirt: with her hair side-parted and held off her forehead with two combs.

Away from the 'labo', Rosalind began to explore her new environment. She went with friends on walking trips to the forest of Chantilly; she found a 'delightfully clean' place to swim in the upper Seine, she took in exhibitions at the Grand Palais, and she mastered the shops. Certain commodities were scarce, but as a *travailleur de force*, she got extra shares of cooking oil, wine and bread, and she gloried in the street markets, full of fresh fruit and vegetables long unseen in London. The maid at the flat tutored her in cooking and she enjoyed improvising: 'Bananas suddenly appeared' and before they disappeared again, almost immediately, 'I had time to invent a pudding – tinned milk, cream cheese, sugar, Fry's chocolate and chopped banana. Excellent.'

Visitors from London began to descend on her – relatives, mainly. Jenifer and Colin were among the first, and she found them a room four minutes away. She was delighted to entertain her mother, who came to visit while the widow was away. Rosalind made up a bed in the drawing room, got tickets to the Comédie Française, organised a visit to an exhibition of Impressionism and prepared meals, which they ate in her room. Her friend Jean Kerslake, now Jean Kerlogue, and pregnant, stayed on the way to Geneva. The formidable Ellis himself came over to Paris for a two-day visit and Rosalind cooked him a dinner too. For all their disagreements, her father was the only one in the family with whom she could talk about her work.

She had a steady list of requests for things to be brought from London. Sweetened tinned milk was top of the list ('if you have points to spare', she told her mother), which included other essentials of the English diet, such as Bovril, marmalade, Fry's drinking chocolate and dehydrated potatoes; also, toothpaste, plain postcards ('an invention unknown to the French who only have them covered with the Eiffel Tower'), a rubber bathing cap, tennis balls and tyres and inner tubes for her bicycle.

Her bicycle was important as she did not like using the Metro. The claustrophobia remarked on by her supervisor in Cambridge and her cousin in Putney made her feel in the Metro that she could not get enough air. By cycling, she argued, she was breathing the pure Paris air which allowed one to work better than in London. However, she knew enough about the Metro to contradict her father: 'As for your remark about Paris "decadence", Paris crowds particularly in the Metro are no worse than before the war. The manners and general behaviour of the police have most strikingly improved and manners in polite society are as perfect as ever.'

Installed in the room next to hers at the Quai Henri IV within a few months of Rosalind's arrival was a young Italian crystallographer recently arrived from Argentina. Vittorio Luzzati, born in Genoa, had emigrated to Buenos Aires at the start of the war. Now in Paris, married to a French doctor who had also been in Argentina, he got along well with Rosalind from the start. He liked her intelligent, expressive face and thought she had good French for an Englishwoman. To him her facility for experiment was remarkable and he quickly saw that she had 'golden hands'. They had a mutual taste for fierce argument and there was plenty to argue about. He was a trained X-ray crystallographer; she was a physical chemist. 'When Vittorio and Rosalind were together,' said an onlooker, 'it was hammer and tongs and quite exhausting.' Rosalind would never let anyone get away with a careless or unsubstantiated statement. Also, as foreigners both proud of their French, they were gleeful in spotting each other's errors. (However, in what was to be a long friendship, they never used anything but the 'vous' form of address to one another.)

One day she asked Luzzati's advice on what to wear to a Christian funeral service. How should he know? He was Jewish. When Rosalind said that she was too, they had a good laugh. Each had been fooled by the other's assimilated surname.

Rosalind all her life felt more comfortable with Jewish people.

By no means were all her close women friends Jewish, but most of her men friends were. While it is hardly rare for any scientist to have Jews as colleagues, it is a fact that, throughout her career, Rosalind's most imaginative and productive research was done with male scientists of Jewish background.

As fate would have it, the expert guide to her introduction to X-ray crystallography was not only Jewish but also the archetypal seductive Frenchman, the deliberate charmer who plays on any woman's susceptibility. '*Toutes les jeunes filles au labo étaient amoureuses de Mering*' ('All the girls in the lab were in love with Mering'), said one who saw him in action. Jacques Mering, with high Tartar cheekbones, green eyes and hair combed rakishly over his bald spot, found in Rosalind the best student he ever had: brilliant, hungry to learn, incredibly dexterous in her research techniques and ingenious in experiment design. At the same time, she was highly attractive to him, with her trim figure, lustrous hair, glowing eyes, and eagerness to catch his every word. Mering, not being a professor, was inexperienced in teaching students in a group and was more comfortable on a one-to-one basis. He and Rosalind would spend whole days and on into the evenings deep in discussion and lively argument over the internal arrangement of atoms in irregular crystals.

Mering was known to be married – that is, he had a wife from somewhere in his pre-war existence, but not in immediate evidence. It was a cruel dilemma for Rosalind: a man who was magnetically attractive and intellectually inspiring, yet who did not see the need for a divorce to sort out his private life, courting adoration all the while.

Rosalind, to the French observers around her, was puritanical, in what seemed the English manner. Adrienne Weill, well-acquainted with Rosalind's background, attributed her emotional immaturity to her family, particularly to her father, but also to her education – to St Paul's, which turned out 'terrible, hearty hockey-playing English girls who are embarrassed to be women' and to Newnham College, which (Weill felt) gave out the message that intellectual women should not expect to combine a family and career.

Rosalind's English friends noticed a change in her appearance on her visits to England – the first occasion perhaps being her brother David's wedding to Myrtle Montefiore in March 1947. Seeing the smart New Look, the make-up, the new hair-do and necklaces, they drew the conclusion that Rosalind was in love. 'Something *happened* to Ros in Paris,' decided her cousin Ursula. Anne Crawford sensed that there might have been 'some sort of sexual involvement'. However, another Cambridge friend ascribed the transformation to scientific success.

From what has come to be known of the principals in this small drama, it seems fair to deduce that Mering was very drawn to Rosalind and made advances of some sort, and that she allowed herself to be tempted farther than was usual for her but eventually, incapable of a casual liaison, drew back and dreamed only of scientific collaboration, of their names linked in papers of their joint work. Mering would later admit there had been 'something' between them, that he loved her very much but that she took things 'seriously' and was naive and inexperienced.*

Did her parents sense danger? Muriel Franklin was quite upset when Rosalind asked if Mering, who was about to visit London and return an overcoat that Roland had left behind in Paris, could stay the night at Pembridge Place. Muriel did not welcome the thought of putting up a French-speaking stranger. At this, Rosalind blew up; she could hardly expect Mering to deliver the coat, she said, and then to go out and look for a hotel. That, in the end, seems to have been what happened.

Rosalind never drew back from her passion for high mountains. In the summer of 1947 (which was extremely hot) she went with

* Rosalind's sister, Jenifer Franklin Glynn, and her brothers, Colin and Roland Franklin, strongly disagree with Anne Sayre's interpretation of Rosalind's relationship with Mering and do not believe that Rosalind had the romantic feelings Sayre attributes to her.

a walking party in the Haute Savoie. After nearly a decade of climbing, she had developed a Wordsworthian capacity to conquer precipitous peaks, walk miles in fog and storm, then come home and write rapturously about it. In a letter to Colin she de scribed a tough sixteen-hour hike along the Peclet-Polset Arêtes and Aiguilles; then, next morning, 'We started out in cloud at 4.30 a.m. and the cloud lifted suddenly at sunrise, just as we came onto the glacier, revealing pink summits above a "mer de nuages". I cannot describe the effects, I can only tell you that the sheer beauty of it made me weep.'

With her fluent French and her knowledge of Paris, Rosalind became a point of call for her family's wide London circle. 'The Cousinhood' could be demanding. In the spring of 1948 she was sought out by one of her new in-laws, Harold Montefiore, and at the same time by her Cambridge friend Rachel Caro, plus Rachel's brother Anthony (later the renowned sculptor). Rosalind organised an evening party that included Rachel's cousin Pierrette who lived in Paris. The evening was not a success. Harold Montefiore turned up 'in black hat, large white silk scarf and white gloves, looking more like the Mad Hatter than anything I've ever seen. It turned out that he and Tony Caro had been bitter enemies together at "prep school" and had succeeded in avoiding one another ever since. And Pierrette doesn't get on with the Caros anyway. However,' Rosalind concluded revealingly, 'we all felt we'd done our duty.' Capable as she was of fierce defence of her rights, she never protested against demands made in the name of family.

Harold's return was followed by another request from the Montefiores, apparently parsimonious although much richer than the Franklins:

> I've just had an absurd telegram signed 'Montefiore' and sent by the Wayfarers! [The Wayfarer's Travel Agency was one of the extended Franklin family's businesses.] I'm very pleased

> to book rooms for any of them but I rather object to being dealt with through a travel agency. And I think they might have paid for one more word to let me know which of them to expect. I shall book expensive, and this side of the river because I haven't the time to go the other side.

Rosalind also continued her intense interest in the plight of Jews driven out of Europe and, although no Zionist, read the papers attentively as the day for the end of the British Mandate and the proclamation of the new state of Israel approached. A birthday letter to her father in late March 1948 answered his request for information on the French press's coverage of Palestine – 'less concerned' than the British, which she found too pro-Arab. 'Who,' she demanded a week later, 'is responsible for the shocking article in *The Economist?*' (*The Economist*'s leader, 'Realities in Palestine', said bluntly that 'There can now be no settlement in Palestine of any kind but force.')

She was far more worried that the developing arms race between the West and the Soviet Union could only lead to another war, and she scolded her father – and his whole generation, who had 'seen the 1914 war and consequences' for having been 'so little cynical about the last war'. She favoured Western European union and thought national sovereignty nonsense.

However, she picked up the BBC on her wireless and asked her parents to substitute a subscription to the *Radio Times* for *The Economist*. She remained faithful to the *New Statesman*. She saw Laurence Olivier's *Henry V* three times, and in at least one other respect, was proud to be British. Seeing France torn between de Gaulle and a very militant Communist left, she diagnosed the weakness of the French left as 'the absence of a strong and efficient socialist party . . . the French are extremely envious of the effective British system of equal sharing of food and the relatively scarce necessities'.

From the very start, she accepted that her main reason for working abroad was that she had not found a suitable job at home.

She regarded her return to London as inevitable – quite apart from the steady drumbeat of letters from her parents asking when she was coming home. (It was only London that she would consider: for her the English provinces or Scotland were unthinkable.) She calculated (correctly) that her employment prospects would improve once she had published more work, which would not be long. Five papers on the density, structure and chemical composition of coals were now either published or accepted for publication. The first which gave her institutional affiliation as the Laboratoire Central, although it was based on her BCURA work, appeared in *Transactions of the Faraday Society* in 1948: 'A Study of the Fine Structure of Carbonaceous Materials.' On her frequent return trips to London she renewed old contacts and visited labs at the Royal Institution and Birkbeck College with an eye out for possible openings.

France, seen from Britain in the late 1940s, was a faraway country and getting there was not half the fun. The Channel crossing, before the ferries were equipped with stabilisers, was an ordeal. Evelyn Waugh's *Vile Bodies* opens with the spectacle of embarking passengers:

> to avert the terrors of sea-sickness they had indulged in every kind of civilized witchcraft . . . some had filled their ears with cotton wool, others wore smoked glasses, while several ate dry captain's biscuits from paper bags . . .

Still:

> Sometimes the ship pitched and sometimes she rolled and sometimes she stood quite still and shivered all over, poised above an abyss of dark water; then she would go swooping down like a scenic railway into a windless hollow and up again with a rush into the gale . . . and sometimes she would drop dead like a lift. It was this last movement that caused the most havoc among the passengers.

By her second year in France Rosalind – almost certainly dipping into her personal money in England – discovered air travel. It was so easy, she announced, she would continue to use it. She had arrived at Heathrow with 'a good five minutes in the waiting room', during which she watched late passengers drive up to the plane in their own cars. Arriving at Le Bourget, having no luggage, she was first through customs. It made a short visit worthwhile.

Even so, she remained utterly absorbed in her X-ray work in Paris. The apparatus, sophisticated for its time, consisted of a demountable X-ray tube evacuated by a double oil and mercury diffusion pump. These demanded considerable maintenance and cleaning. Michel Oberlin, a young recruit to the lab in October 1948, recalled seeing her first crouched on all fours, reassembling a vacuum system. It was a messy job; litres of benzene were needed to free the pumps of sludge and clean off vacuum grease but she did not mind. Another day, she saw that he noticed dirt on her face and explained, 'It's not dirt. It's graphite.' As she worked, neither she nor her colleagues were overly concerned with the dangers of radiation. In comparison with later tight safety standards about shielding and protection, safety procedures were lax, and the staff were comparatively insouciant about the dangers. Jacques Maire, a research student who was assigned to her, caused a permanent injury to his finger through his haste and indifference to safety while processing X-ray film in the darkroom; decades later, he shrugged it off as a small matter. Rosalind, for her part, was very annoyed when her radiation monitoring badge showed that she had exceeded the permitted levels and had to stay away from the lab for several weeks.

The marks of the post-war world went far beyond the political. In the United States the publication of the Kinsey Report told an astonished world that sexual behaviour was more frequent and varied than allowed in conventional morality or the Hollywood Film Code. The publication in Paris of Simone de Beauvoir's *Le Deuxieme Sexe* signalled that women would no longer accept the role of bystanders at life's pageant. And Cambridge University quietly joined the trend by officially recognising women as members of the university; it bestowed degrees retroactively on women students such as Rosalind who had graduated before 1948.

Rosalind's stay in France was not the unalloyed bliss that some who visited her there believed. A dark moment came in the summer of 1948 when a group from *'la bande de Solange'* went together to Corsica for a holiday. Their destination was the port of Calvi, and the Hotel di Fango, a small pension owned by Guastala, a friend of Adrienne Weill's, and his former wife, an architect, who had built it. Requisitioned by the Germans during the war and dilapidated, the semi-ruin was open to selected friends who were prepared to rough it. There were six couples in the party and they agreed to take turns to do the cooking.

The composition of the group might have signalled trouble, for one of the 'couples' was herself and her brother Colin. Colin soon realised he had been asked along because Rosalind was the only one of that group who had no partner: 'Nor, come to that, had I. Except for Roland, we were in that way a boring frightened lot.' Rosalind had, however, made herself a bikini (just coming into vogue) for the occasion.

She painted a happy picture for their parents in her holiday letter home. After a hectic week in Paris when she got through a piece of work she wanted to finish, she had left 'utterly exhausted. Now after two night journeys and two late nights dancing I'm completely refreshed – Calvi is by far the most beautiful place I've seen.'

That much was true. So was her information that on the night boat to Calvi she had travelled fourth-class and slept out on the deck. She did not mention, however, and perhaps did not notice, what much amused her French companions – that she settled her sleeping bag right outside the door of the cabin being shared by Mering and Rachel Glaeser, another of the young women in the lab. Naively, Rosalind had not realised, although everybody else in the lab knew, that Mering and Rachel were having an affair. The lovers were much amused by Rosalind curling up outside their door. It reinforced their opinion that Rosalind 'was like Queen Victoria about men'.

If so, the sight of unmarried couples openly cohabiting brought her into the mid-twentieth century. Before long, she and Colin went off by themselves to explore the island. When they returned, Colin was joined by Roland and a young man from Oxford whom Rosalind found 'most unpleasant to me and my friends'. The three non-scientists formed 'a sort of in-group' sitting at the edge of Rosalind's circle of scientists and before long they went off to Italy. She was thus doubly isolated.

Back in Paris Luzzati's wife, Denise, observed that Rosalind was, in her controlled way, very upset about the liaison. Yet the whole episode had a curious consequence. Rosalind accepted Rachel as a mother figure and Mering's permanent partner (as she would remain for the rest of his life). Rather, she transferred her hostility to Agnès Mathieu-Sicand, another pretty young woman in the lab who appeared to have caught Mering's roving eye. With her capacity for intense antagonism, Rosalind burned with jealousy of Agnès for the rest of her time in Paris. A telling photograph of this combustible quintet, taken outside the Ecole de Physique et Chimie, shows Rosalind, stylish in a floral print with square shoulders and pearls, and Agnès, more flashily dressed, closely flanking Mering, while Michel Oberlin (who later married Agnès) and Rachel look happily unconcerned.

Paris did bring Rosalind a new English-speaking female friend. Anne Sayre was a writer and lawyer who appeared at the lab one day to collect her husband David, an American crystallographer who had just finished a stint in Oxford. Rosalind introduced herself and asked if the couple needed help settling in to Paris. To Anne (who remained an enormous admirer), Rosalind looked 'very crisp and very pretty' in a cotton print in white and lilac. Anne was struck by 'bright inquisitive eyes and a beautiful flashing smile'; also, by a shy manner and a lovely speaking voice: 'English as she should be spoke'. The two women became good friends and Anne sensed that there were undercurrents of strong feeling at the lab. But she never pressed for specific details and none were volunteered.

Rosalind was beginning to make an international reputation. At meetings in Nancy, Lyons and Paris, she met scientists from Britain, the United States and the Continent. Attending one conference in Paris simply as a member of the audience, she found that her BCURA work suddenly became relevant. She rose to her feet and, in her good French, summarised her findings. The chairman, noticing the respectful silence around the room, asked her to repeat her contribution in English. After that she found herself invited, like an official delegate, on a tour of Versailles and the Petit Trianon and treated to the kind of tea with sweet cakes which she loved.

Photographs taken on these occasions show her in the white-shirt, dark-skirt combination adopted as business uniform by professional women in the era before trouser suits became acceptable for women in the workplace. Yet at Lyons in 1947, obviously enjoying herself, she lightened her look with open-toed platform sandals, an Italian-style beach bag and bare legs, combining self-confidence with fashion consciousness and femininity.

She stepped up her efforts to return to England. The problem was to find the right job. She implored her parents, 'I wish you wouldn't appeal in every letter to come home for a long holiday. I can't be in two places at once, and at the present I'm here, but not forever.' In the spring of 1949 after two years abroad, she reiterated to her mother, that 'it would be a *bad* idea to come home without a job. I might well hang around for a year and nothing could be more depressing. I took a year from the time I started to look for a job from Cura to the time I got here.' She had thought of working at the Royal Institution in London but ruled it out when, at one of the Paris meetings, she met the man (whom she did not name) likely to supervise her research and found him 'odious'. The only remaining possibility was Birkbeck College's research laboratory, but thought her chances would be better of finding the sort of job she wanted in England once she had more papers to her name.

In the autumn of 1949, she was refused the job she had applied for at Birkbeck, and, as she told her parents, 'I can't face the man at the Royal Institution – so I suppose I'm here for a bit longer. Anyhow my work here is going extremely well and it would be a pity to drop it at the moment. Getting a job should be easier when it is published.'

Birkbeck now boasted Bernal on its staff, he having shifted from Cambridge after the war with the intention of using X-ray crystallography to extend physical chemistry into biology. Rosalind was not the only Birkbeck reject that year. Also turned down was a physicist, Francis Crick, who, leaving work with the Admiralty in London, wished to join Bernal. Crick was told by Bernal's secretary, 'Don't you realise that people from all over the world want to come and work under the Professor?' Instead, Crick found himself at the Strangeways Laboratory in Cambridge, and before long, at the Cavendish Laboratory, where his path and Rosalind's would cross.

For her twenty-ninth birthday Rosalind asked for a subscription, at £2 10s, to the new journal established with Unesco support, *Acta Crystallographica*. *Acta Cryst*, as it is called in the field, was soon so deluged with manuscripts that it increased its annual issues from six to eight. Rosalind's interest was not only as a reader but as a contributor. Her detailed X-ray investigation of carbons and graphites yielded results that in June 1949 were sent off to *Acta Cryst*. What would be her ninth published paper, 'The Interpretation of Diffuse X-ray Diagrams of Carbon', was written under her name alone. Mering, for reasons best known to himself but possibly related to her determination to return to England – he did not like apostates and was a man who bore grudges – withheld his name from what was in many ways their joint work. Luzzati saw her close to tears in disappointment.

Even so, she was gaining confidence and independence. That summer her work was going so well that she drafted a letter (the 'letter' was the accepted format for a brief paper) to the eminent British scientific journal, *Nature*. She sent a copy as well to Professor Charles Coulson of King's College London, a theoretical physicist whose paper was cited and whom she knew from her BCURA days. Coulson found her conclusions in 'On the Influence of Bonding Electrons on the Scattering of X-rays by Carbon' to be 'entirely reasonable'. He made some helpful comments, and regretted that he would be away in late June when she was coming to London but hoped to see her again. The *Nature* letter was accepted, after she made some modifications suggested by referees, and was scheduled for publication in January 1950. She was now in the big league of working scientists.

The harder she worked, the harder she walked. Holidays became more important than ever to her (if that were possible) and she planned these with the meticulousness of her lab work. Through a friend at Unesco, she enlisted a young Australian woman as a companion on a walking tour in the Italian Alps in the summer of 1949. Margaret Nance, a scientist recently arrived from Australia to do a PhD at the Lister Institute of Preventive Medicine in London, was eager to enjoy any chance to explore any part of Europe and accepted the loan of the ice axe Rosalind borrowed for her. They climbed, with a guide, the peaks above Aosta. 'Rosalind certainly planned it all and I did what I was told,' said Margaret.

Rosalind's planning extended to her own clothes. Her kit included not only a tailored anorak and hiking boots, but a stylish patterned jersey turban of the kind popularised by Coco Chanel, which she wrapped around her head to protect her ears.

Holidays with her brothers were now finished. David Franklin had married in 1947. In 1948 Roland, at twenty-two, married Nina Stoutsker, the daughter of a cantor at the Central Synagogue, London. In 1949 Colin became engaged to Charlotte Hanjal-Konyi, whom he had met at Oxford (and whose family also claimed descent from the great Rabbi of Prague) and they married in 1950. Nannie Griffiths came for every wedding. There is no hint that any relative or guest at these ceremonies was tempted to ask the eligible unmarried elder sister when they might expect to go to hers.

All three sons had by now married within the faith, as would Jenifer in time. Colin, looking back, said, 'My father would never have countenanced any of his five kids marrying a non-Jew.' However, others who knew Rosalind well are adamant that had she found the right man, his religion would not have mattered to her nor would have her father's opposition. But somehow she did not.

Colin had gone up to St John's College after the war. When he took his degree, having absolutely no taste for banking, he accepted the paternal nudge into Routledge's, the family publishing firm.

With Colin leaving the nest, Ellis and Muriel Franklin decided it was time to leave 5 Pembridge Place for a smaller house in north London. Rosalind's own long-term plans, she wrote Colin, asking what he wanted for a wedding present, were as vague as ever. All that was certain was that, as her Paris work continued satisfactorily, 'I shan't come home until I have a job.'

SEVEN

Seine v. Strand

(1950)

EARLY IN 1950 Rosalind began reading the job advertisements in *Nature* and, more specifically, asked Professor Charles Coulson at King's College London how to apply for one of the Imperial Chemical Industries research fellowships available at a number of British universities. Coulson explained that the procedure was to find an appropriate department and get the consent of the departmental head before applying. He steered her in the direction of King's College London, where in 1947, reluctantly surrendering the comforts of Oxford, he had become professor of theoretical chemistry with the possibility of venturing into biology. 'If you are interested in possible biological applications of the technique that you now know so well,' Coulson wrote her in February, 'there could be quite a lot to be said in favour of King's.'

Rosalind did not know biology; she had not taken even gardening-level botany at St Paul's but she was willing to learn: 'I am, of course, most ignorant about all things biological,' she replied, 'but I imagine most X-ray people start that way.'

Far from preparing to leave Paris, however, after three years in the widow's garret, she organised a proper flat of her own. Rented accommodation was almost non-existent, as housing controls gave priority to French citizens, who then acquired security of tenure. Rosalind, however, succeeded in finding a furnished flat owned by a French commandant stationed in Germany, who was willing to let to foreign *locataires* because he would not risk losing control of his property. The apartment was a set of balconied rooms just under the mansard roof at 16 Avenue de la Motte-Picquet in the seventh arrondissement, near the Ecole Militaire. To share the cost, she found flatmates: an American couple, Philip and Marion Hemily. He was an American Fulbright scholar working under Vittorio Luzzati for a University of Paris doctorate in molecular structure determination by X-ray diffraction analysis. His wife Marion, with a bachelor's degree in physics, was in the American government's Paris office for implementing the Marshall Plan. Although the new apartment was further from the lab, it offered independence, with no great comfort, at a more elegant address.

The place was crammed with dusty Louis XV furniture; the heating came from a coal-fired boiler in the kitchen. Philip, an American war veteran charmed by his new life, was impressed by Rosalind' s knowledge of Parisian ways, such as ordering coal from the local café and having it delivered to the basement of their address. Every day Philip carried the bags of '*boulettes*' up the six flights of stairs.

Rosalind found herself in a triangle, this time a tense one. She did not get along with Marion Hemily. The living arrangements – two women and one tiny kitchen, with Rosalind's bedroom next to the shared salon – almost guaranteed friction. Rosalind, moreover, disapproved of the Hemilys' habit of sleeping late. She got along far better with Philip. She enjoyed cycling with him, especially if Marion did not come along, into the countryside at weekends.

At the Labo Central, Rosalind was responsible for overseeing a small group studying the X-ray diffraction of carbons. Papers poured out. Her work began to gain wide notice because of the great industrial potential of the kind of carbon, known as 'glassy' or 'vitreous', that did not form graphite upon heating. Bell Telephone Laboratories, among others, were interested in the hard carbons that could be made into crucibles, tubes and other articles requiring reliable heat resistance.

Her debut in *Acta Crystallographica* was made in June, the paper sent the previous year and her most important so far: twelve pages, with nine diagrams and many mathematical proofs, summarising one of her main findings: that some carbons cannot be transformed into graphite no matter how much you heat them.

At the end of the paper, in the list of acknowledgements which were eagerly scanned by readers, she thanked Marcel Mathieu and J. Desmaroux, head of the Labo Central, for providing facilities, and, in conclusion, 'Monsieur J. Mering, for his guidance and very generous help throughout the work.'

In March 1950, with work pressing at the lab, she treated herself to a flying visit to London to look up Coulson at King's College. He introduced her to the head of King's physics and biophysics, Professor J.T. Randall. With Randall's endorsement, she put in an application for an ICI fellowship, telling her parents, 'half of me hoping I don't get it'. It was not a particularly distinguished award, nor did it lead in any specific direction.

Back in Paris, she went off with the Luzzatis to Italy. Her holidays, in which so much hope was invested, rarely disappointed her. This one, she proclaimed once more, was 'the best I ever took'. She loved lying around for three hours in the middle of the day and found the countryside around Florence so beautiful they abandoned the idea of going to Rome. (As she became closer to the Luzzatis, Rosalind drew back from Rachel Glaeser and Jacques Mering. Denise Luzzati could see the depth of Rosalind's feeling for Mering and felt there was a hopelessness about it which Rosalind accepted.)

Rosalind was interviewed in London by the ICI and Turner and Newall Research Fellowships Committee early in June 1950, and by mid-month she had a fellowship for three years, starting in the autumn, to work at King's College under J.T. Randall. There she would work, according to the Fellowships secretary, 'by means of X-ray diffraction, on proteins in solution and the changes in structure which accompany the denaturing [heating or dehydrating] proteins in solution'.

Flattered and disappointed – 'I'd really like 1 more year here' – she applied for a postponement to the beginning of the next year, 1951, which was granted. She then was off with the Luzzatis to Normandy.

In mid-June Randall reported her appointment to the principal of King's College, saying that the Turner and Newall Committee had placed her 'pretty well top of the list' and that her work in applying X-ray diffraction to the structure of various microcrystalline carbons was well known to Coulson and himself. The money was more than decent for the times: £750 a year for three years. (In 1950 junior lecturers at British universities were paid about £400 a year.) Rosalind was not pleased, however, when her cousin Ursula's husband, Noel Richley, a journalist with Reuters, informed her that Turner and Newall was a large North Country manufacturer of cement and asbestos. She had long prided herself on staying aloof from industry.

No sooner had Rosalind secured her fellowship when the direction of research at King's College changed dramatically. In May, at a meeting of the Faraday Society (whose *Transactions* had carried three of Rosalind's papers on coal), a Swiss scientist named Rudolf Signer from Berne displayed a specially prepared gel of nucleic acid which had extremely high molecular weight. Signer generously offered samples to his audience. One of the takers was Maurice Wilkins of King's College physics department, who took his back for study.

That hardly concerned Rosalind. Rather, she was overwhelmed with doubt now that her return to England was settled. She told Colin that her feelings 'still oscillate wildly'. 'When I left London I thought I'd convinced myself it wasn't too bad but cycling along the Seine I was miserable'. She was touched that her personal assistant was heart-broken.

As its name implies, the Strand, the busy thoroughfare that runs east from Trafalgar Square, once was a river bank. King's College London faces the Strand on the north and the Thames on the south. As she prepared to move, the alliterative combination of 'Seine' and 'Strand' stuck in Rosalind's mind, and she used it, sometimes intensifying the sound with 'cellar', in letter after letter: 'to change the banks of the Seine for a cellar in The Strand seems to me quite insane'.

Over and over, she cited reasons for dreading return. Britain's nuclear deterrent was high on the list; she was appalled at her country's willingness to side with the Americans. One of the things which made her feel more comfortable in France than in England was in France 'there is a much healthier horror of war, and appreciation of its senselessness – but you won't appreciate that – nobody who hasn't lived outside England ever does'.

Even so, she cast her mind to the prospects of finding a place to live in London, where housing was even harder to find than in Paris. Holidaying once more with the Luzzatis in the Haute Savoie in August (her third with them that year), she overheard English people talking about a flat in London near Gloucester Road in South Kensington. Eavesdropping eagerly, she gathered two details: the flat had a very low rent of £6 per month but high key money of £750. She wrote her parents immediately for advice: 'I have no idea how immoral and illegal such deals are in England or what guarantee that low will stay low.'

On that Alpine trip, the Luzzatis noticed how adept Rosalind was at keeping men at arm's length. In a mountain hut one day, they met a climber who, to them, clearly fancied Rosalind. Chatting her up, the stranger mentioned that he had lost a button off his coat; could she help him? 'Of course,' Rosalind said, in her brisk, well-bred way, searched in her rucksack, found a needle and thread and handed them over for him to do the job himself. They were amused too, when at a large Luzzati family gathering, Rosalind was a bit shocked by all the kissing. She never kissed any of her family, she said, except her mother.

By the end of October, with her return to London approaching, Ellis and Muriel urged her, as parents will, to use her family connections to help get the South Kensington flat. She protested. 'I can't possibly write to Lord Somebody or Other and say I'm Mamie's niece and I want a flat . . .' (As a member of the London County Council, 'Mamie' – Helen Bentwich – was a woman of influence.) Rosalind reminded her parents that she was by no means sure she wanted to come back: 'I spend half my time wondering whether to chuck the whole thing up and stay here.'

Her gloom at the prospect of re-entry was deepening. She poured out her complaints about England to Evi Ellis. Rosalind felt she could count on the Austrian-born Evi to understand her dislike of English parochialism. After making the Seine-Strand contrast, she warmed to her theme: 'What depresses me most in the English is their vacant stupid faces and child-like complacency – I'm busy collecting the addresses of foreigners in London.' She begged Evi to write and tell her how to get out of going back.

In dealing with King's College, however, she was very far from trying to get out of her new job. She put a great deal of thought into a long, respectful and detailed letter on 24 November to Randall, specifying, with a long numbered list, the equipment she would need. She was particularly concerned with the design of the camera being made for her. What was essential was to control the temperature inside the camera to prevent the solution from changing during its exposure. An economical move, she suggested, would be to put the specimen *outside* the vacuum cover, and only the camera inside. (A vacuum was necessary because X-rays are attenuated when they travel through air.) She offered some cost figures to suggest that the apparatus could be designed more cheaply and quickly in Paris, allowing her to get down to work as soon as she reached London. Then too she would need the new X-ray generating tube promised by Birkbeck College. If she could know when the order would be placed, she might ask for a few minor modifications.

In conclusion, she apologised to her new boss: 'Please forgive me for bothering you with so many questions after such a long silence.'

Ten days later Randall replied. He was changing completely the direction of her planned research. His artful and ambiguous letter was packed with half-truths and buried meanings that would explode in Rosalind's face before too long and, not incidentally, alter the course of scientific history. It began:

> After very careful consideration and discussion with the senior people concerned, it now seems that it would be a good deal more important for you to investigate the structure of certain biological fibres in which we are interested, both by low and high angle diffraction, rather than to continue with the original project of work on solutions as the major one.

Randall then rattled off without explanation references to people of whom she may or may not have heard, including 'Dr Stokes who now wished to concern himself almost entirely with theoretical problems'. Then came the critical sentence:

> This means that as far as the experimental X-ray effort is concerned there will be at the moment only yourself and Gosling, together with the temporary assistance of a graduate from Syracuse, Mrs Heller.
>
> Gosling, working in conjunction with Wilkins, has already found that fibres of desoxyribose nucleic acid derived from material provided by Professor Signer of Berne give remarkably good fibre diagrams.

Having told her that she would have the X-ray work to herself, Randall held out the possibility that she might look at proteins in solutions at some future time but, for the moment, 'we do feel that the work on fibres would be more immediately profitable and, perhaps, fundamental'.

If she was startled by this change of assignment, Rosalind gave no sign. Paris had been good for her. She left it at Christmas 1950, an authority on the structure of coal and carbons, and thinking that perhaps she should consider staying in France indefinitely. She reminded her parents that she liked Europe and the Europeans so much better than England and the English; Italy and Italians would suit her just as well as the French and she had always liked 'foreigners better than the English'. She dreamed of a job that would keep her half the year in Paris and the rest in London.

She summed up her regrets in a strong letter to Colin and his new wife Charlotte (whom she liked very much), first thanking them for a copy of *Christ Stopped at Eboli* which she had read in Italian. Even her favourite brother, however, cannot have caught the subtle reference to the narrow and unsatisfactory relationship that told her she had no future in Paris:

> . . . I still cannot believe that I'm leaving here next January – I can see no way out of it now, but I'm sure it was the biggest mistake of my life. At times I think seriously of just saying that I've changed my mind and will stay here. But at this stage it would mean upsetting such a lot of people that it's not worth doing unless I really decide to stay here indefinitely – and I'm just not quite sure enough for that. If only I had managed to meet more of the right people here and had a wider circle I shouldn't hesitate. Apart from that I far prefer here the place, the people, the life and the climate. I feel – and felt even before I came to France – far more European than English. National feeling, whether it be for England or any other country, is meaningless to me. However, I suppose my fate is decided and I ought to stop thinking about staying on here – though that is impossible. I am doomed to spend the next 3 months moaning about the future and a good many months after that moaning about the present.

Her intimation of trouble ahead was well-founded but she was very far from making the biggest mistake of her life. J.T. Randall, however, had just made his.

Part Two

EIGHT

What Is Life?

THE FIRST HALF OF twentieth-century science belonged to physics, with the general theory of relativity, quantum mechanics and nuclear fission. The second half would belong to biology. In the post-war world, the secret of the gene – how hereditary characteristics pass from one generation to another – was the hottest topic in science.

For a number of physicists who had worked on the Manhattan Project to develop the atomic bomb, the post-war shift into biology was a stark exchange of the science of death for the science of life. But their conversion was as much intellectual as ideological. Biology was now where the action lay. The war had interrupted a line of investigation leading towards understanding the chemical basis of heredity.

That physical features are passed on by discrete units (later called genes) had been discovered in 1865 by the Austrian monk Gregor Mendel in his experiments with garden peas. Each gene determined a single characteristic, such as height or colour, in the next generation of plant. By 1905 it had been learned that within living cells the genes are strung together like beads on the chromosomes, which copy themselves and separate. But how does the genetic information get from the old chromosome to the new?

Protein was the obvious candidate. By the 1920s it was thought that genes were made of protein. The other main ingredient in the chromosome is deoxyribonucleic acid, or DNA. DNA, a substance of high molecular weight, was identified in 1871 by a young Swiss scientist, Friedrich Miescher. (There is, in fact, a second kind of nucleic acid in the cell, called RNA, with a slightly different chemical composition.) The 'D' in DNA stands for 'deoxy' – a prefix often spelled as 'des' in Rosalind's day, a usage now obsolete – which identifies it as the ribonucleic acid with one fewer hydroxyl group. But as RNA exists in cells mainly outside the nucleus, it was unlikely to be the genetic vehicle.

Protein was far more interesting to geneticists than DNA because there was a lot more of it and also because each protein molecule is a long chain of chemicals of which twenty kinds occur in living things. DNA, in contrast, contains only four kinds of the repeating units called nucleotides. Hence it seemed too simple to carry the complex instructions required to specify the distinct form of each of the infinite variety of cells that constitute living matter.

In 1936, at the Rockefeller Institute on the Upper East Side of Manhattan, a microbiologist called Oswald Avery wondered aloud if the 'transforming principle' – that is, the carrier of the genetic information from old chromosomes to new – might not be the nucleic acid, DNA. No one took much notice. DNA seemed just a boring binding agent for the protein in the cell.

During the pre-war years, in Britain, J.D. Bernal at Cambridge and William Astbury at Leeds, both crystallographers, began using X-rays to determine the structure of molecules in crystals. Astbury, interested in very large biological molecules, had taken hundreds of X-ray diffraction pictures of fibres prepared from DNA. From the diffraction patterns obtained, Astbury tried building a model of DNA. With metal plates and rods, he put together a Meccano-like model suggesting how DNA's components – bases, sugars, phosphates – might fit together. Astbury concluded – correctly, as it turned out – that the bases lay flat, stacked on each other like a pile of pennies spaced 3.4 Ångströms apart. This '3.4 Å' was no gratuitous detail. Published with other measurements in an Astbury paper in *Nature* in 1938, it was to remain constant throughout all the attempts to solve DNA's structure that were to come.

But Astbury made serious errors, his work was tentative, and he had no clear idea of the way forward. By the time of the Second World War, no one knew that genes were composed entirely of DNA.

In 1943 Avery, at sixty-seven, was too old for military service. Still working at the Rockefeller Institute and building on an experiment with pneumococcus (bacteria that cause pneumonia) done by the English physician Frederick Griffith in 1928, he made a revolutionary discovery. He found that when DNA was transferred from a dead strain of pneumoccous to a living strain, it brought with it the hereditary attributes of the donor.

Was the 'transforming principle' so simple then – purely DNA? In science, where many grab for glory, there are some who thrust glory from them. Avery, a shy bachelor who wore a pince-nez, was one of those too modest for his own good. His discovery has been called worth two Nobel prizes, but he never got even one – perhaps because, rather than rushing into print, he put his findings in a letter to his brother Roy, a medical bacteriologist at Vanderbilt University Medical School in Nashville. 'I have not published anything about it – indeed have discussed it only with a few,' he said, 'because I am not yet convinced that we have (as yet) sufficient evidence.'

A year later, however, Avery, with two colleagues, wrote out their research. In what became a classic paper, they described an intricate series of experiments using the two forms of pneumococcus, virulent and non-virulent. When they freed a purified form of DNA from heat-killed virulent pneumoccocus bacteria and injected it into a live, non-virulent strain, they found that it produced a permanent heritable change in the DNA of the recipient cells. Thus the fact was established – at least for the readers of *The Journal of Experimental Medicine* – that the nucleic acid DNA and not the protein was the genetic message-carrier.

The essential mystery remained. How could a monotonous substance such as DNA, like an alphabet with only four letters, convey enough specific information to produce the enormous variety of living things, from daisies to dinosaurs? The answer must lie in the way the molecule was put together. Avery and his co-authors, Colin MacLeod and Maclyn McCarty could say no more than that 'nucleic acids must be regarded as possessing biological specificity the chemical basis of which is as yet undetermined'.

In 1943, another scientist at one remove from the world conflict (because he had been offered a haven in neutral Ireland) gave a series of lectures in Dublin, called provocatively 'What Is Life?' An audience of 400 for every lecture suggested that his supposedly difficult subject was of great general interest.

Erwin Schrödinger, a Viennese, had shared the Nobel prize in physics in 1933 for laying the foundations of wave mechanics. That same year he left Berlin where he had been working because, although not himself Jewish, he would not remain in Germany when persecution of the Jews became national policy. A long odyssey through Europe brought him, in 1940, to Dublin at the invitation of Eamon de Valera, Ireland's premier. De Valera had been a mathematician before he became a revolutionary, then a politician; in 1940 he set up the Dublin Institute of Advanced Studies. Schrödinger found Ireland 'paradise', not least because it allowed him the detachment to think about a very big question.

In his Dublin lectures, Schrödinger addressed what puzzled many students – why biology was treated as a subject completely separate from physics and chemistry: frogs, fruit flies and cells on one side, atoms and molecules, electricity and magnetism, on the other. The time had come, Schrödinger declared from his Irish platform, to think of living organisms in terms of their molecular and atomic structure. There was no great divide between the living and non-living; they all obey the same laws of physics and chemistry.

He put a physicist's question to biology. If entropy is (according to the second law of thermodynamics) things falling apart, the natural disintegration of order into disorder, why don't genes decay? Why are they instead passed intact from generation to generation?

He gave his own answer. 'Life' is matter that is doing something. The technical term is metabolism – 'eating, drinking, breathing, assimilating, replicating, avoiding entropy'. To Schrödinger, life could be defined as 'negative entropy' – something *not* falling into chaos and approaching 'the dangerous state of maximum entropy, which is death'. Genes preserve their structure because the chromosome that carries them is an irregular crystal. The arrangement of units within the crystal constitutes the hereditary code.

The lectures were published as a book the following year, ready for physicists to read as the war ended and they looked for new frontiers to explore. To the molecular biologist and scientific historian Gunther Stent of the University of California at Berkeley, *What Is Life?* was the *Uncle Tom's Cabin* of biology – a small book that started a revolution. For post-war physicists, suffering from professional malaise, 'When one of the inventors of quantum mechanics [could] ask "What is life?"' Stent declared, 'they were confronted with a fundamental problem worthy of their mettle.' Biological problems could now be tackled with their own language, physics.

Research into the new field of biophysics inched forward in the late 1940s. In 1949 another Austrian refugee scientist, Erwin Chargaff working at the Columbia College of Physicians and Surgeons in New York was one of the very few who took Avery's results to heart and changed his research programme in consequence. He analysed the proportions of the four bases of DNA and found a curious correspondence. The numbers of molecules present of the two bases, adenine and guanine, called purines were always equal to the total amount of thymine and cytosine, the other two bases, called pyrimidines. This neat ratio, found in all forms of DNA, cried out for explanation, but Chargaff could not think what it might be.

That is where things stood when Rosalind Franklin arrived at King's College London on 5 January 1951. Leaving coal research to work on DNA, moving from the crystal structure of inanimate substances to that of biological molecules, she had crossed the border between non-living and living. Coal does not make more coal, but genes make more genes.

NINE

Joining the Circus

(January–May 1951)

A CELLAR IN THE STRAND was an unlikely place to look for the secret of life. But the subterranean setting was all too appropriate for what has been called one of the great personal quarrels in the history of science.

In 1951 the contrast between Paris, which had escaped the attentions of the Luftwaffe, and London which had not, was all too apparent. The main quadrangle at King's College London, stretched between the Aldwych and the Thames Embankment, was occupied by a bomb crater 58 feet long and 27 feet deep. The biophysics unit in the basement wound round the pit, which was being excavated to build the new physics department. The South Bank seen from across the Thames was simply piles of rubble. London was depressing, at its bleakest since the outbreak of war, with bomb sites used as car parks, cracked buildings propped up by wooden buttresses and makeshift housing everywhere. Six years after the war's end, food was still rationed, and sallow faces showed the effects of years of privation. Menus offering 'meat and two veg' delivered a plate of beige mutton and two forms of overcooked potato. The Welfare State that Rosalind had so admired from France was fraying round the edges. An economic crisis brought on by the onset of the Korean War had, among other things, ended the free spectacles and false teeth offered by the new National Health Service, and in the general election of February 1950 the Labour government's post-war majority of 146 fell to a mere 5.

The contrasting genders often attributed to Paris and London were visibly true. London was a man's city, a place of furled umbrellas, bowler hats, men's clubs and gentlemen's tailors: to the expatriate writer Jean Rhys, it was 'patriarchy personified'. So it was for Rosalind. In Paris she was confident, admired, independent. Now she was a daughter again. At first she lived at home with her parents at their new home in north London, then moved to the flat in South Kensington secured by Ellis Franklin's business connections. Dutifully but not unwillingly, she crossed London to her parents' Friday night dinners where the arguments flared as before. Rosalind's intolerance of her father's conservative views, her mother decided, stemmed from associates in France who had been in the French Resistance, 'many of them politically embittered Communists'.

Rosalind was thus in a state of glum apprehension when a graduate student at King's, whose wife had worked with her at BCURA, volunteered to show her around the college. John Bradley found her tense and unbending, reluctant to be introduced. Clutching her aversion to things English around her like a cloak, she looked for, and found, mediocrity.

One element in English life unmentioned in her foreboding letters from France was class. At the start of the 1950s the social divisions were almost as pre-war. The Angry Young Men had not yet been heard from. Not until 1956 would *My Fair Lady* inform a popular audience that 'every time an Englishman speaks, he makes another Englishman despise him'. Rosalind's quick, clipped voice carried a class label.

The impression she made on Dr Jean Hanson, for example, the biophysics unit's senior biologist was 'typically upper-class – a product of the best girls' schools':

> if you are English you feel it about another person. I always supposed her family was rich, though she never talked about it – she really stood out very much around here where most of the other people with very few exceptions came from ordinary backgrounds, middle-class or in some cases, I sup-

> pose, lower than that. Really, the word is aristocratic – she looked like an aristocrat and she acted like one . . . just the way she spoke, there were people at that time who sneered at the upper-class way of speaking, and really *hated* it.

The accent divide cut both ways. King's, part of the ill-coordinated University of London, was no Oxbridge college. Its corridors echoed with something less than Received Pronunciation. To Rosalind, the voices signalled that she had got herself among the intellectually second rate.

There was nothing second rate about King's tradition in laboratory science. The college had appointed its first professor of natural and experimental philosophy in 1831, long before Oxford established the Clarendon Laboratory or Cambridge the Cavendish. In 1834 Charles Wheatstone, King's first professor of experimental philosophy, laid down iron and copper wires in the vaults for experiments in electricity. In 1860, James Clerk Maxwell was appointed professor of natural philosophy and proved that electricity and magnetism are aspects of the same phenomenon.

But science at King's had always breathed ecclesiastical air. Theology was its biggest department. With 400 students training for the Anglican priesthood, there were swirling cassocks and dog collars everywhere. The college was established in 1829 by the Church of England as a protest against the opening of University College London, the 'Godless Institute of Gower Street'. UCL, founded by dissenters, was denied a royal charter because it admitted Roman Catholics, Jews and other non-Anglicans. King's, in conspicuous contrast, boasted a large chapel with vaulted ceiling (and later designs by George Gilbert Scott), the only consecrated building in the University of London. The sound of hymns from morning service rang in the ears of the scientists as they made their way along the flagged passages and down the stone stairways, past a door marked 'Professor of Dogmatic Theology', to their labs in the basement.

Rosalind soon was informed that women were not allowed in the King's senior common room where some of the staff ate lunch.

With happy memories of the Labo Central's disputatious *déjeuners* at Chez Solange, she felt angry and excluded. It seemed as if her work was not going to be taken seriously. But she ought not to have been surprised. It was hardly a unique arrangement in London at the time, certainly not in a bastion of the Established Church. Women were still not employed at Keyser's bank. Even free-thinking University College, the first to admit women with full status, had one common room for men only, and a separate, joint, common room – for men and women; known familiarly as 'the joint', it survived well into the 1960s; UCL women, when polled, chose to retain the status quo.

Like UCL, King's had two dining rooms, one for men and women, the other for men only, both served from the same kitchen. Many of the men preferred to eat in the communal dining room, overlooking the Thames, and some of the scientists refused to go at all into the male preserve because of the preponderance of 'hooded crows' (clerics).

There was yet another element at King's foreign to Rosalind. The biophysics unit included a number of ex-military men who had come to King's for an intensive two-year undergraduate course, and who remained on, working together as a team. They had a tough, ferocious approach to work and play – a barracks-room, beer-drinking camaraderie that spilled over after hours into Finch's pub on the Strand.

In short, an upper-middle-class Anglo-Jewish woman with French tastes in serious discourse suddenly found herself in an environment friendly to everything she was not. Adjustment was not helped by a mournful postcard from the Faculté des Sciences de l'Université de Paris, Laboratoire de Minéralogie: '*Chère Mademoiselle, Nous regrettons tous à Paris votre départ.*'

On 8 January 1951, J.T. Randall, FRS, called a meeting in his office to introduce Rosalind and discuss the way ahead. Present were three of the people he had mentioned in his letter changing Rosalind's direction of research. Most important to her was

Raymond Gosling, a young doctoral student, the only one at King's using X-ray diffraction, and now assigned to assist her. Gosling joked that he was 'the PhD slave boy handed over in chains'.

Gosling liked jokes. Young, endomorphic, good-natured, he referred to Randall as 'JT', 'the Old Man' or 'King John'. Nonetheless, he saw the boss as a visionary genius and considered himself lucky, a product of University College Medical School, to get into Randall's sought-after biophysics unit. Gosling had heard that Rosalind was pretty high-powered. Now he could see that she had 'beautiful dark eyes, shining black hair, an intensity about her, and an awkwardness in conversation'.

Also present was the physicist and mathematician Alec Stokes who Randall had said wished 'to concern himself almost entirely with theoretical problems'. There too was the 'graduate from Syracuse, Mrs Heller'. Louise Heller was working as a volunteer at King's while her husband studied in London on a Fulbright scholarship. She too formed a strong impression of the new arrival: 'very attractive, very bright, very impatient and very opinionated'.

More important was who was not there. Maurice Wilkins, Randall's deputy, assistant director of the biophysics unit, was on holiday. Wilkins had been working intensively on nucleic acids at King's for several years. If Gosling was anybody's slave, it was his. Together they had mounted a bundle of nucleic acid fibres on a wire frame (the 'frame' being a bent paperclip stuck in a holder), then kept the sample moist by passing saturated hydrogen through their Raymax tube and Unicam camera. When warned by Randall that air might leak into the camera (spoiling the vacuum necessary to prevent the X-ray beam diffusing), they sealed it with a condom. The result was what Randall had described to Rosalind in the crucial letter:

> Gosling, working in conjunction with Wilkins, has already found that fibres of desoxyribose nucleic acid derived from material provided by Professor Signer of Berne give remarkably good fibre diagrams.

Gosling's photographs were more than remarkable, they were unique. The spots forming an x-shape were the first clear indication that deoxyribonucleic acid – DNA – had a crystalline structure. When Wilkins showed Gosling's measurement of the spots to Stokes, what struck Stokes were the blank spaces along the length of the 'x'. Might the molecule, he speculated, have the shape of a helix?

Wilkins believed he was instrumental in getting Rosalind assigned to DNA. When he heard from Randall that she was coming to work on proteins in solution, he thought it a waste as they were getting such good results on nucleic acids. Considering her X-ray expertise, why not, he suggested, 'grab her and get her in on the DNA work'? To his surprise, Randall readily agreed. Randall usually hemmed and hawed over such decisions. With new equipment on the way, Maurice had looked forward to her joining his team.

King's had been using an X-ray machine loaned by the British Admiralty. When the Admiralty wanted it back, Wilkins and Gosling went to Birkbeck and asked to buy the new fine-focus tube that Werner Ehrenberg and Walter Spear had invented to concentrate X-rays into a very narrow beam. Ehrenberg generously gave them the prototype instead. This tube, when fitted with a very small camera, made it easier to control the humidity inside the camera. The way was now open to photographing a single DNA fibre one-tenth of a millimetre across.

Rosalind's first task was to complete and try out the new apparatus and to order more. With brisk know-how and an array of smart-looking catalogues, she ordered equipment from Paris, adeptly haggling in French about price and specifications, and an English-made vacuum pump to extract air from the camera. With Gosling she set about designing a tilting microcamera to be made by King's workshops.

At the same time Randall encouraged her to write up her Paris work. Within a few weeks an extensive paper was on its way to the *Proceedings of the Royal Society,* with his endorsement: 'communicated by J.T. Randall, FRS'. Publication in this distinguished journal was another first for Rosalind. It gave her the opportunity to expound on the subject on which she was now a world authority: the two distinct classes of carbons and the way that each formed crystals. With two more papers on their way to publication in *Acta Crystallographica,* she had, at the age of thirty, an impressive representation in what the scientific profession respects as 'peer-reviewed journals'.

If anyone at King's was class-conscious, it was John Turton Randall. A short, bald, dapper Lancastrian, the son of a market gardener, he was a grammar-school boy who had worked his way from a county school to the University of Manchester where he led his year and took a first-class degree in crystal physics. When, somewhat to his dismay, he was steered by the Nobel Laureate Lawrence Bragg, then head of physics at Manchester, into industry rather than pursuit of a doctorate, he rose to become head of research at the General Electric Company based in Wembley. Yet he remained self-conscious about the 'rough edges' of his northern ways and the difficulties in becoming 'fully acceptable in the smoother south'. In 1937 he was given a fellowship and invited to the University of Birmingham where he set up a luminescence laboratory. In 1939 he shifted to the problem of increasing the power of radar.

Randall was – and knew he was – something of a war hero. His invention (with H.A.H. Boot) of the cavity magnetron, a device to create a powerful beam of low frequency electromagnetic waves which made possible the detection of distant objects. This invention was an invaluable aid to bombing at night and in overcast conditions, and was critically important in hunting down German submarines in the Battle of the Atlantic. President Franklin Roosevelt called the cavity magnetron 'the most valuable cargo ever to reach these shores'.

It was certainly valuable for Randall. Well before the war

was won, he put in his claim to the Royal Commission on Rewards for Inventors for inventions which had served the national good. He pursued his bounty assiduously over the next few years as he moved from Birmingham to St Andrews in Scotland, where he held the chair of Natural Philosophy. He even managed to secure tax relief and compensation for legal expenses incurred before finally collecting, in 1949, his one-third share of £36,000. (J. Sayres, another Birmingham researcher, and H.A.H. Boot each also got a third.) Randall thus became financially independent beyond dreams of ordinary academics.

The biophysics unit at King's could be seen as another form of reward from a grateful nation. Randall wished, post-war, to apply the techniques of physics to large molecules of biological importance. Endorsement for the unit's formation came from the Royal Society, but the actual money was provided by the Medical Research Council. The MRC, which was to loom large in his life, encouraged him to move down from remote St Andrews to the livelier environment of King's College London, where he set about gathering his staff. For his project, the MRC gave him £22,000 – a generous nest-egg much resented by some others in other King's departments, struggling under limited budgets.

One of Randall's first recruits was Maurice Wilkins, a Cambridge graduate, who had been with him at St Andrews and Birmingham. Wilkins had spent much of the war in Berkeley, California at the Livermore Laboratory, working on the Manhattan Project. After the war, he read Schrödinger's *What Is Life?* and moved his interest to the molecules controlling living processes. For Wilkins the shift to biophysics was a change from a dry science to a wet one. When he recoiled from one of Randall's experiments – cooling horse blood in liquid air to obtain an absorption spectrum – Randall reminded him that the great Michael Faraday had once put a beef steak between the poles of his magnet.

At the only physics department in Britain to have a major research interest in biophysics, Randall and Wilkins were the king and crown prince, but Randall wore a triple crown. He was at one and the same time: Wheatstone Professor of Physics, head of the physics department, and honorary director of the Medical Research Council biophysics research unit within the physics department. Randall was good – shockingly good, some felt – at snaring people on all sorts of grants and fellowships, which got his motley assembly of talent the name 'Randall's Circus'. He was less good at delineating their responsibilities once he got them. He preferred to communicate by notes on pink or white paper, and he was manipulative, not above playing people off against each other.

A bit of a showman, he wore dandyish American-made suits, a bow tie and a fresh flower in his buttonhole every day and cherished a rubber plant kept in his office. (It was suggested that his interest in applying physics to biology derived from the love of plants acquired in his father's nursery.) At the same time, he was, in Wilkins's view, 'tough as old leather – a Napoleon who sat at a big desk and liked to see people falter as they approached it'.

Randall was unusual for his time for the prominence he gave to women scientists. When Rosalind came to King's, eight out of thirty-one of the biophysics staff were women, several in quite high positions. Dr Honor Fell (later Dame) was Senior Biological Adviser to the unit, coming once a week from Cambridge where she was director of the Strangeways Laboratory. Dr Jean Hanson (who became FRS in 1967), was working on muscle and took the main responsibility for the biological side of the work. Dr Marjorie M'Ewen, a lecturer in physics, was another recruit from St Andrews. At least half a dozen others were active in research, not to mention the laboratory photographer Freda Ticehurst, whom Randall had recruited personally from his old lab at GEC, and who worshipped the ground he walked on.

In the early 1950s women engaged in scientific research were still a rarity, regarded much as Dr Johnson, two centuries earlier, living in a house not far from King's, dismissed a woman's preaching as like 'a dog's walking on its hind legs. It is not done well; but you are surprised to find it done at all.' Women scientists almost never won a Nobel prize: at that time there had been only three, two of them Curies, mother and daughter. The Royal Society had by 1951 admitted only seven as Fellows – less than 1 per cent of the total. (The percentage would remain below four for the rest of the century.)

Of all the sciences, moreover, physics was, as it has remained, the most male-dominated. The science historian Margaret Wertheim in 1995 dubbed it 'the priesthood of science'. In her interpretation, the persistent cultural and psychological barriers to the entry of women into physics are a legacy of ancient religious tradition: the physicist or mathematician was a kind of priest, a conduit to God the divine mathematician. Physics departments in the egalitarian United States were scarcely more welcoming to women than they were in Europe. Harvard University's physics department in the 1950s, maintained a policy against the hiring of women even as instructors – a ban that endured for a further two decades. (No woman professor gained tenure in physics until 1992.) Princeton was worse. In the 1950s not only were women forbidden to teach physics, they were not allowed into the physics building. They were, the head of the department believed, a distraction. A female nuclear physicist invited to Princeton to use the cyclotron had to creep in under cover of darkness.

Randall's lab offered a strong contrast to this misogyny. Not only did he have many women on his staff, but they tended to find him as an employer sympathetic and helpful. Himself married, with an invalid wife, Randall liked women around him, and rumours persist of a personal attachment to one of his staff. He also enjoyed social life, and liked to see his flock, men and women, come together for morning coffee, and at lunch in the joint dining room where he ate with them nearly every day. Above all, there was afternoon tea: 'a command performance', Louise Heller recalled. Another command performance was the annual departmental cricket match, with Randall, a cricket fiend, donning flannels and pads to bat for biophysics.

When Maurice Wilkins got back to King's in January 1951, he got his first look at Rosalind. She was sitting at a desk in a small office and when she turned round he, like Gosling, was first struck by her eyes – 'steady watchful dark eyes'. He found her 'quietly handsome' and exuding such confidence that when she stood up, he was surprised to see that she was not as tall as he had thought.

During the next few months they worked on DNA separately but cordially. She wrote up her Paris work for what seemed to him a long time but he did not bother her. He continued to look through microscopes to watch how DNA fibres swelled and lengthened as they absorbed water. He discussed with her a paper he was preparing on extensible molecules and included her in his acknowledgements.

He was wary of her peremptory manner. Once when he stood at the doorway talking, without turning in her chair she indicated with a wave of her hand that he should come in and sit down. Yet they occasionally lunched together on a Saturday at the Strand Palace Hotel, where those who had come into the lab – mainly those who were single – went for the buffet, which was inexpensive and good. Conversation was not difficult. Rosalind was interested in theatre, books and politics. Like him, she was a *New Statesman* devotee and, with Britain joining the North Atlantic Treaty Organisation, she advocated neutrality in the Cold War. Both of them disliked intensely the idea of science for profit. Occasionally what Wilkins called 'spikiness' would show. One Saturday as they ate fruit salad with cream, he remarked that the cream had been good. 'But it was not real cream,' Rosalind corrected him. Wilkins felt rebuked. Not having lived in Paris in the post-war years, he had obviously forgotten what real cream tasted like.

* * *

In mid-century, despite O.T. Avery's discovery at the Rockefeller Institute in 1944, it was by no means generally accepted that the nucleic acid DNA was the genetic material. Many scientists still believed that genes were made of protein. Wilkins, for his part, had picked on the molecular structure of the gene for its intrinsic interest, seeing it as the most rewarding problem to attack as he moved into biology. Before long, however, he and his colleagues received important advice from a colleague who had worked in New York: 'trust Avery'.

By 1951 King's had made good progress towards figuring out the structure of DNA. Randall said so himself. In a lecture to the Royal Society on 1 February, he began by asking that his new unit not be 'judged too severely' in comparison with much longer-established institutions, alluding to its apparently slow start.

In his talk, titled 'An Experiment in Biophysics', Randall outlined with some pride a varied programme of investigation into the nature of the cell by all optical techniques, by a research staff of twenty-six, including one Turner and Newall Fellow (i.e., Rosalind). In discussing his unit's advances in nucleic acids, he generously gave credit to work done in London at Birkbeck. Sven Furberg, a young Norwegian researcher there on a British Council scholarship, had woken them all up to an error in the pre-war X-ray work of Astbury of Leeds. In a model he had built of the nucleic acid, Furberg showed that the sugars were at right angles to – not parallel with – the bases stacked like pennies.

At the same time Randall paid tribute also to the sample of DNA they were working on at King's – the gel supplied by Professor Signer of Berne. When Wilkins and others had wetted this gel, they had drawn fibres that could be extended to double their original length – a process called 'necking' – and then shrunk back again.

Randall alerted his audience to the audacity of his attempt to map the living cell. The congestion and disorder of the King's College site, he said, was a good symbol for biophysics: 'This breaking down of the mental and physical boundaries associated with workers of highly different training and tradition is a necessary beginning to successful research in a borderline field.'

The words of a pompous pioneer, perhaps, but what was the point of modesty? Had he not helped win the war?

Rosalind was working without benefit of academic appointment or rank. Like a distinguished refugee arriving in a new country without baggage or reputation, she had entered a field in which the quality of her previous work was largely unknown except to the two professors, Randall and Coulson. Wilkins did not appreciate how senior she had been in Paris. Her other colleagues certainly did not. Holes in coal, or the lack thereof, were of little interest to a biophysicist.

The people she found most congenial were two women junior to herself: a young Austrian, Marianne Friedlander, and Freda Ticehurst, who ran the photographic laboratory which developed and printed the X-ray films the scientists took. Freda, an excellent technician, was warm, fun, and motherly. She was also mothered by everybody, for she had lost her fiancé during the war. Randall certainly did when, in September 1950, on the pretext of borrowing some slides, he had returned to his old GEC lab at Wembley, and invited her to come to King's to equip and run the photographic laboratory.

Freda's darkroom was a refuge for the entire lab. Randall himself often stopped by for a chat. Rosalind too would come in for a break, particularly on the first day of her menstrual period when she suffered from severe cramps. She would be given an aspirin and a hot drink. But sympathy was not what she wanted. Her reaction was 'Oh well. I just have to put up with it.' Freda sympathised with Ray Gosling too, at those times of Rosalind's month.

Rosalind's craving for foreign friends in London was rewarded when she met Simon and Bocha Altmann from Argentina. Simon Altmann, a Jewish refugee to South America during the war, was a friend of Vittorio Luzzati's; they had been students together at the University of Buenos Aires. Altmann had come to King's from Argentina as a graduate student in theoretical physics; his wife Bocha, a biochemist, was a British Council fellow at University College.

Altmann was appalled to see the predicament in which Rosalind found herself. Coulson, his own professor in theoretical physics, himself a reluctant recruit from Oxford (to which he would soon return), kept his distance from the macho rowdies in 'Experimental'. To Altmann, it was little short of a tragedy that someone of Rosalind's sensitivity and ability should be in an atmosphere so utterly alien to her. 'Very well read in two languages, she was used to a civilised intellectual life, discussing painting, poetry, theatre and existentialism,' he said. Now she found herself among people who had never heard of Sartre, whose chief reading was the *Evening Standard*, and who enjoyed 'the type of girls that would get drunk at departmental parties and be passed from lap to lap having their bra undone'. People at the lab knew about Rosalind's friendship with Luzzati and thought he was her lover. Altmann knew he was not.

Altmann felt that Rosalind, with five years of distinguished post-doctoral work to her credit, was undervalued by all at King's, not least by Randall. The personal antipathy between her and Wilkins was just a part of the problem.

The bright spot in Rosalind's London life was her new flat on Drayton Gardens in South Kensington. Modestly furnished, it looked luxurious to her friends most of whom lived in cramped quarters with shared bathrooms or kitchens. Four rooms on the fourth floor of a purpose-built, 1930s block, Donovan Court, it was around the corner from the Fulham Road and the Forum Cinema, and was approached by a lift and a quiet, wide hallway. Windows at the back looked out at the streetlamps of the paved walkway, Thistle Grove.

Rosalind loved the flat, the first real home of her own. She made her own curtains and pressed friends to use the place when she was away. Her American friend from Paris, Anne Sayre, claimed credit for persuading Rosalind to dip into her personal capital for her own comfort, abandoning her usual self-denying ordinance to live only on her wages. Penny-pinching had made some sense in Paris, Anne argued, when currency restrictions forbade the import of funds. But for anyone who was not genuinely hard-up (as Anne felt she and her husband were), it was self-indulgence to economise from choice rather than necessity. To Anne's amazement, Rosalind 'not only took my lecture meekly but proceeded to take the flat, and to use – though always sparingly – the private income she had always previously scorned'.

Just as she was a serious scientist and serious walker, Rosalind was a serious cook. A decent kitchen all to herself gave her the opportunity to entertain as had not been possible before. She gave many dinner parties and, cooking being a branch of chemistry, was very good at it. She introduced her English guests to French cuisine with the same missionary air with which Elizabeth David's highly popular book *Mediterranean Cooking* was preaching the virtues of olive oil, garlic, parmesan and basil to enliven the dreary national diet.

Rosalind served her London friends unaccustomed foods such as pigeon, rabbit and artichokes. She also instructed them in the arts of the French kitchen. How to tell whether a Camembert is ripe? Press one finger on the cheese and the other to the closed eyelid; if the consistency matches, the cheese is ready. Potatoes taste better cooked without water. Put them in butter in a heavy pan, cover and cook them in their own juices. She adored putting garlic in things, especially when her father came to dinner. Ellis Franklin who always insisted he hated it never recognised it in his favourite dish, roast beef.

Those lucky enough to be invited for a meal met the other Rosalind: not the one from the lab where she was 'someone who kept to herself' and who never smiled, but rather the amusing conversationalist and attentive hostess. Not so much Jekyll and

Hyde as *sol y sombra*. To an extraordinary degree Rosalind compartmentalised parts of herself. Those who saw one side rarely saw the other.

Her guests often included members of her family, such as her aunt Alice Franklin who lived nearby at Elm Park Gardens and to whom Rosalind was close – perhaps, her cousin Ursula thought, because Alice did not get on with Ellis. 'Is it your turn or mine to be favourite niece this week?' she would ask Ursula. The Altmanns were frequent guests; younger members of the biophysics unit were also favoured with a meal: Louise Heller and her husband, Raymond Gosling, and Freda Ticehurst who could see Gosling's awkward position, caught between Rosalind and Wilkins.

Of all the 'what might have beens' surrounding the quest for the structure of DNA, Rosalind's failure to invite Maurice Wilkins to dinner at her flat deserves a high place on the list.

Rosalind's great-grandparents, Ellis Abraham Franklin and Adelaide Samuel Franklin.

Her grandparents: Arthur Ellis Franklin and Caroline Jacob Franklin.

Her great-uncle (*centre*), Sir (later Viscount) Herbert Samuel, first British High Commissioner to Palestine.

Three sons and a daughter: Ellis Franklin with his children on a beach holiday, ca 1926.

A smile to come home to: Nannie Ada Griffiths, the centre of the Franklin children's lives.

Norland Place School Golden Jubilee: Rosalind, wearing spectacles, is seated in the front row.

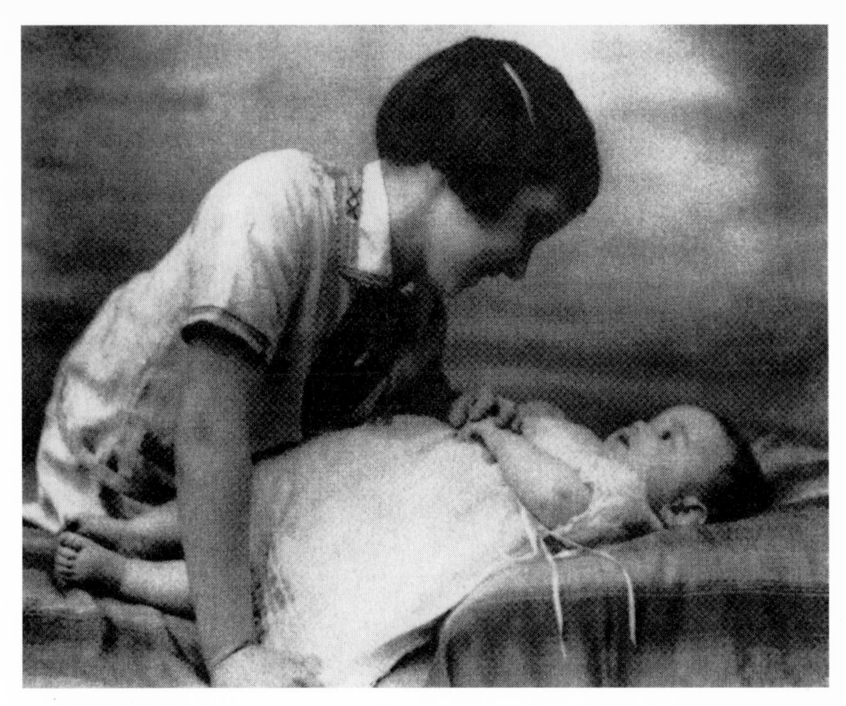

Rosalind and her new sister, Jenifer.

Rosalind (*back row, far right*), as a hockey-playing Paulina.

Right Rosalind in Norway with her precious climbing boots.

Below Ellis A. Franklin, Rosalind's father and sparring partner.

Below right Working scientist: Rosalind with colleagues at the British Coal Utilisation Research Association (BCURA), 1945.

Opposite Five shots of Rosalind on the mountain holidays she loved.

Rosalind with friends in Brittany.

Bare-legged Rosalind with colleagues in Lyons, 1949.

Philip H. Emmett, Rosalind – wearing glasses – and Marcel Mathieu at the Congrès de catalyse de Lyons, 1949.

At the same conference in Lyons: (*left–right*), Haisinski, Jacques Mering, Rosalind (beaming at her mentor), Irene Perrin.

Outside l'Ecole de Physique et Chimie in Paris: (*left–right*) Rosalind, Mering (holding Rosalind's handbag), Agnes Oberlin, Michel Oberlin and Rachel Glaeser.

Adrienne Weill enjoying a joke with Mering.

Vittorio Luzzati's snapshot of Rosalind in the Cabane des Evettes on a mountain holiday, 1950/51.

The after-lunch coffee-making ritual at the 'labo'.

Above Rosalind, following Luzzati, at Uppsala on an excursion to an island in the Stockholm archipelago during the 1951 International Union of Crystallography Conference at Stockholm.

Left Professor J.T. Randall.

Below Interdepartmental cricket match, King's College London: (*left–right*) unidentified, Maurice Wilkins, Bill Seeds, Bruce Fraser, Mary Fraser, Raymond Gosling (standing) and Geoffrey Brown.

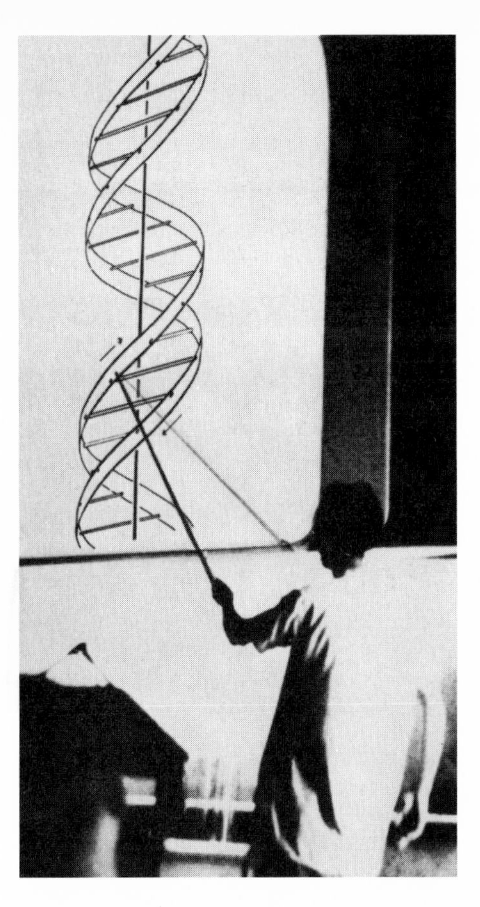

James Watson at Cold Spring Harbor, Long Island, summer 1953, demonstrating the double helix he and Crick had just discovered.

Maurice Wilkins.

Below Rosalind's lab at Birkbeck on the top floor of 21 Torrington Square.

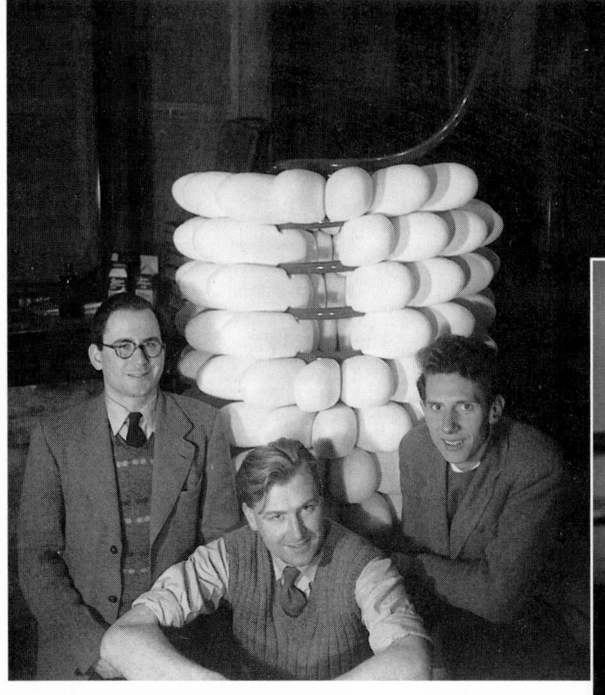

Rosalind's young team at Birkbeck: James Watt, John Finch and Kenneth C. Holmes.

Professor J.D. 'Sage' Bernal.

Isidore Fankuchen, Dorothy Hodgkin, J.D. Bernal, and Dina Fankuchen, September 1939.

Madrid, 2 April 1956, at an International Union of Crystallography Symposium on 'Structures on a scale between the atomic and microscopic dimensions': *(left–right)* Francis Crick, Don Caspar, Aaron Klug, Rosalind Franklin, Odile Crick and John Kendrew.

Right Rosalind's last look at the Matterhorn, 15 July 1957, with Don Caspar and Richard Franklin (no relation).

Muriel and Ellis Franklin on their ruby wedding anniversary in August 1957. Despite her illness Rosalind organised the celebration; shortly afterwards she was rushed to hospital.

Opposite, top Watson, Crick and their DNA model in 1953. For purposes of the picture, Crick was persuaded to point with a slide rule even though he had not used one in preparing the model.

Below To the victors: the Nobel prize line-up in Stockholm, December 1962, four years after Rosalind's death: Maurice Wilkins, Max Perutz, Francis Crick, John Steinbeck, James Watson and John Kendrew.

31
THOMSON
SANDLER

Too deep for tears: Rosalind in a pensive mood.

TEN

Such a Funny Lab
(May–December 1951)

WHEN SPRING COMES, says Chaucer, 'Thanne longen folk to goon pilgrimages.' In science these are called conferences. Especially in the early 1950s, with currency restrictions severe and salaries small, the chance to attend, expenses paid, a conference abroad was a conspicuous sign of status and carried a glamour hard to recapture in a jet-weary age. At that time the Medical Research Council's annual report listed the foreign visits made by members of its research establishments during the preceding year as if these were scientific achievements in themselves. In May 1951, Wilkins thus was happy to stand in for Randall at a conference on large molecules being held at the Marine Biological Station in Naples.

At Naples, Wilkins spelled out very plainly for his international audience the reason for the concentration on nucleic acid at King's. When living matter is prepared in crystal form, he said, the arrangement of its molecules can be seen and may lead to an understanding of the structure of the gene. He then put a slide on the screen. No one had ever shown such a sharp discrete set of reflections from the DNA molecule. There was nothing like it in the literature. William Astbury from Leeds, who was in the audience, congratulated Wilkins at the end for this expansion of his own studies. Astbury, an unpersuasive man working at an unfashionable northern university, had not been able to match Randall's post-war knack of attracting government research money.

Also in the audience at Naples was Dr James Watson – the 'Dr' being particularly remarkable as he was only twenty-three. He had entered the University of Chicago in 1943 at the age of fifteen. When he was seventeen, in the university biology library, he too had read Schrödinger's *What Is Life?*, and changed his own life. He determined to learn what the gene was. Reorienting himself to genetics, he went for his doctoral studies to the University of Indiana; here the Italian microbiologist Salvador Luria brought him into the elite circle under the legendary Max Delbrück to study bacterial viruses. This line of research won Watson a postdoctoral fellowship to Copenhagen to work with the biochemist Herman Kalckar.

As he had heard of neither J.T. Randall nor Maurice Wilkins, young Watson was not disturbed by the common phenomenon of a last-minute replacement for a well-known name on a conference programme. The subject – nucleic acids – was what had drawn him to Naples. He listened eagerly to Wilkins's report that the nucleic acid, DNA, could be prepared in crystal form (as not all large molecules can). Crystalline DNA, as Wilkins was arguing, then could be subjected to X-ray diffraction and thus its three-dimensional structure might be deduced.

Watson believed – with a single-mindedness unavailable to the war-weary heads around him – that the gene was the thing to study. However, he feared that it might be, in his words, 'fantastically irregular' in shape. The clear pattern of Wilkins's slide demonstrated to him there was a regular structure waiting to be mapped. (Genes should not be equated with DNA; they are *made* of DNA, but not all DNA molecules are genes.) By understanding the structure of the molecule DNA, Watson hoped to understand how the gene did its work of replication.

Watson was a self-propelling young man who had achieved a modicum of adolescent fame on *The Quiz Kids*, a national radio show. His protruding eyes gave the look of being ready to pounce, and he pounced on Wilkins. Very conscious of his own bachelor state, he gathered that Wilkins was an unattached male.

When he saw his pretty sister Elizabeth, who had come to Naples with him, eating lunch with Wilkins, Watson formed the thought that he might use his sister as a lure. In his words, 'if Maurice really liked my sister, it was inevitable that I would become closely associated with his X-ray work on DNA'.

But where would that work go next? In the middle of 1951 Sven Furberg's thesis was being widely read at King's. Rosalind had a photographic copy of the pages describing Furberg's models and the arrangement of DNA's sugars, bases and phosphates (the cluster of chemical groups that together is called a nucleotide). Young Furberg's model for DNA was in the shape of a helix, and demonstrated, in the words of Harry Carlisle, head of crystallography at Birkbeck, 'that the helix could be a natural structural system for biological macro-molecules'. She was thus acquainted, as were her colleagues, that in the chemical groups inside DNA, the sugars were not parallel to the bases but at right angles to them, but that Astbury had been right in stating that the bases were stacked parallel to each other, 3.4 Ångströms apart.

Furberg, having deposited five copies of his thesis at the University of London Library, returned to Norway, full of admiration for Britain which 'after having suffered, fought and won the terrible war, felt able to do great things also in peace, to fight and win new battles. Huge social reforms were introduced and their scientists formed ambitious goals, like finding the structures of the fundamental molecules of life.'

This was Randall's philosophy in a nutshell. With the same brilliance with which Britain won the war, the boffins would now solve the mysteries of the living cell.

Raymond Gosling, the 'slave boy', found Rosalind 'super' to work for. She took great delight in handling her materials. The golden hands had survived the Channel crossing. She began reassembling the X-ray equipment with the Ehrenberg-Spear tube against which a single fibre would be positioned for long hours of X-ray exposure. The procedure, incredibly laborious, now done in seconds by a computer, was not a job for the ham-fisted.

Next she tackled a problem that Wilkins had encountered – making the humidity inside the camera more stable. She knew very well from her French work the importance of keeping specimens moist during the procedure. Working now with a camera with very small volume made this easier. She chose a series of salt solutions through which to bubble hydrogen into the camera at controlled humidities. She first pulled the water out by placing the DNA fibre over a drying agent, then put it back at will by increasing the humidity to a range of different values. It was sometimes possible to use the same fibre several times. Thus she, with Gosling's assistance, demonstrated the easy, reversible hydration of DNA.

Something else Rosalind appeared to have brought from France was an apparent unconcern with the radiation emanating from X-rays. Neither she nor Gosling wore protective lead aprons, even though they worked with the X-ray beam turned on. The camera could only be aligned when the beam was on. Geoffrey Brown, who was working on the separation of nucleic acids by chromatography, came into the basement X-ray room below the level of the Thames, looking for Gosling one night and found Rosalind in semi-darkness working on the X-ray camera. 'She shouldn't get in the beam,' he thought, and sensed that she was so keen to get the pictures that she was reckless. Louise Heller, too, worried about Rosalind's failure to take proper precautions to save herself from exposure. Even so, Heller, who had worked in health physics at the US atomic energy facility at Oak Ridge, Tennessee, dared not mention it to a woman who 'had the sort of drive that the work was more important than anything else'.

(All of these recollections come with the reminder that scientific experiments in the early 1950s were conducted in a laissez-faire atmosphere of unshielded machines and lax safety procedures unthinkable in a later age.)

* * *

Rosalind's success in using salt solutions to adjust the humidity of the hydrogen-filled camera impressed Wilkins. He had feared the salt might spray the DNA sample. Her technique was the right one. This modest achievement, for her, only increased her scorn for Wilkins. He didn't know simple chemist's techniques of hydrating fibres. Clearly she was not going to have a mentor like Mering, to whom she had just made another genuflection. Her latest paper for *Acta Cryst* on the structure of graphitic carbons thanked 'Monsieur J. Mering', and him alone, 'for his continued interest and frequent advice during the course of this work'.

Wilkins sensed that things were going wrong. He asked everybody what he should do to improve relations with Rosalind. At Gosling's suggestion, he bought her chocolates, but these did not help.

The two should have got along well. Wilkins was thirty-five, tall, gauntly handsome with fine features and long straight hair swept back from his broad forehead. He was gentle in manner and attractive to women. Born in New Zealand of Anglo-Irish parents, he had a gift for metaphor befitting a man whose grandfather had taught the poet W.B. Yeats at Dublin High School. He was mathematically fluent and immersed in the very problems that concerned Rosalind. He came from a liberal Unitarian scientific tradition which included a dedication to the higher education of women. (His grandmother had been one of the first students at Newnham, an aunt had helped establish Bedford College, the first British university to grant diplomas to women, and an uncle was memorialised in the Wilkins Prize in Mathematics for women at Trinity College Dublin.) What is more, by curious coincidence, his own first name was taken from the surname of Frederick Denison Maurice, founder of her father's cherished Working Men's College. Politically, he and Rosalind were in tune. The ingredients for many a laboratory romance were there. That was on the one hand.

On the other, Wilkins was Rosalind's temperamental opposite. Her speech came fast, his slow. He evaded gaze and tended to take off his glasses when talking as if he did not wish to see too much, gradually turning away so that the listener was left facing the back of his head. Rosalind fixed her steady eyes like X-rays on the human specimen before her. She positively liked hot and heavy debate and found arguments with French shopkeepers fun. Wilkins, in the face of conflict, became expressionless and quiet.

If Rosalind had wished, she could have twisted Wilkins around her little finger. Many, then and since, have speculated that he was half in love with her. But just emerged from an abruptly broken marriage – his American wife had refused to exchange San Francisco for St Andrews and sought a divorce, cutting him off from his young son – he was emotionally bruised.

This very eligibility may have put Rosalind on edge. She was far more comfortable with men who were married or much younger than herself. But he had the wrong woman. A charm offensive was not in Rosalind's repertoire. Anyway, she could respect only men who were strong and decisive, with something to teach her. Still under the spell of the charismatic Mering, seasoned by long combat with her aggressive father, she looked at her designated collaborator and found him unworthy.

The Maurice Wilkins who inspired affection and admiration in later generations of students at King's was a man much changed by a happy second marriage and warm family life. The earlier Wilkins could give a different impression. A fellow undergraduate at St John's College, Cambridge, remembered him (with some prejudice perhaps as he himself knew Rosalind and had married a friend of hers) as 'a rather peevish, slightly old-maidy young man . . . none too well off, and with a very large chip on his shoulder – rather a lone wolf . . . and with a passionate interest in optical lenses . . . In many ways, there was much in common with Rosalind but the social backgrounds were so different.'

That seems to have been how he struck Rosalind too. 'He's

so middle-class, Vittorio!' she complained to her old friend when she saw him.

In June it was Rosalind's turn to go on the conference circuit. Well before she left Paris, she had been looking forward to the Second International Conference of Crystallography in Stockholm in June 1951. On the boat over she shared a cabin with the eminent crystallographer, Dorothy Hodgkin FRS, who was violently seasick. Rosalind was not.

Snapshots taken on a day's outing show a happy Rosalind, in her element. Arms crossed, she is wearing a smart shirtwaister dress, trotting down a riverside path in a beautiful European wilderness, in the company of good friends: David and Anne Sayre, Luzzati, and her old flatmate, Philip Hemily were there. (To them she unburdened her troubles, principally that she could not abide Maurice Wilkins.)

It is a conference cliché that the formal sessions are less valuable than the social contacts. At dinner, at coffee breaks, on the inevitable excursion without which, like the group photograph, no conference is complete, and in the bar, scientists tell each other what they are doing – or, according to the microbiologist François Jacob of the Pasteur Institute, 'at least what one wants to leak out and let people believe one is doing'.

Yet Rosalind filled pages with notes about what she learned relevant to her work. The foremost chemist of the time was there: Linus Pauling from the California Institute of Technology, whose classic *Nature of the Chemical Bond* she had read at Cambridge.

Only that spring Pauling had published news of his triumphant discovery – the alpha helix, the most important regular structure found in proteins, enables its internal chains to turn corners. The best showman in science, Pauling had first revealed his idea in a lecture at Caltech when, building up suspense, he waited until the very end to unveil a construction of coloured plastic balls wired together to form a coiled spiral. This, he announced, was the alpha helix's shape.

The news devastated Lawrence Bragg, now director of the Cavendish. Pauling had been his main rival for twenty-three years (in part because of Bragg's suspicion that his own ideas on the chemical bond had been appropriated and passed off as Pauling's own). Now Pauling had solved the structure of one of protein's basic units – a problem that the Cavendish had had in its sights. Bragg slowly walked up the stairs at the Cavendish, downcast, contemplating what he came to call 'the biggest mistake of my scientific career'.

In any event, said Francis Crick, now working on protein at the Cavendish and who witnessed the sorry spectacle, 'Helices were in the air.' Also in the air was the signal that post-war science was becoming competitive, a game of winners and losers.

Pauling's paper, written with his colleague Robert Corey, had been published only two months before the Stockholm meeting in the April and May issues of *Proceedings of the National Academy of Sciences*. At Stockholm he was describing his discovery to an international audience for the first time.

Another speaker at the Stockholm conference was the acclaimed J.D. Bernal from Birkbeck. One of the first to turn X-ray diffraction methods onto giant molecules such as proteins, he addressed, with obvious reference to Pauling, the alternative approaches to working out molecular structures: speculative, deductive and inductive. Whichever was followed – and he did not suggest which was best – the resulting proposal had to be put to the test of X-ray analysis. The important thing, declared Bernal, was to review the evidence and the assumptions, then to ask oneself: 'Have we found *the* solution? Or *a* solution?'

Rosalind took this commandment to heart. She wrote it down with strong underlinings: '*the* solution or *a* solution'. She did not manage to meet Pauling personally.

A less formal conference was held in July at Bragg's Cavendish Laboratory in Cambridge. Max Perutz called, as he had the year before, a meeting of those working on X-ray analysis of protein structure. The Cavendish had a research unit, similar to that at King's and financed by the same source, the Medical Research Council. It was smaller than Randall's outfit and was studying protein by X-ray diffraction. Perutz, a brilliant and witty Austrian transplanted to Cambridge in 1936, was trying to solve the dauntingly intricate structure of haemoglobin, the major protein in red blood cells. His colleague John Kendrew was trying to do the same with the smaller protein, myoglobin ('myo' means muscle).

By the mid-summer of 1951 the atmosphere in Britain had brightened considerably. The Festival of Britain opened, designed to lift the country out of the doldrums: 'the people giving themselves a pat on the back', said the Labour politician Herbert Morrison. Set up on London's South Bank, the exhibition was a £12 million, five-month expression of nationalist faith in the British imagination to conquer new worlds – in art, in architecture and, not least, in science. J.T. Randall and Sven Furberg were not alone in believing that Britain's victory against terrible odds sprang from its scientific genius.

At Perutz's conference, Wilkins spoke, as he had at a similar meeting the previous year, about the DNA work at King's. All the different X-ray patterns that they were getting, he said, showed a clear central x. Alec Stokes had informed him that this was a strong sign of a helix. It seemed, Wilkins concluded, that all DNA had the same unique, twisting structure. He was applauded.

But not by everyone. Rosalind was waiting for him as he left the hall. She told him, in firm and deliberate tones, to return to his optical studies – in essence, to give up X-ray work. Her actual words as he remembered them, were 'Go back to your microscopes.'

Wilkins was profoundly shaken. No one in science had ever talked to him like that. And why was she ordering him, the assistant director of the lab, to stop work just as he reported encouraging progress? He wondered if the success of his talk had disturbed her.

The rift was Randall's doing. Rosalind assumed that Stokes

and Wilkins were moving off DNA, whereas Wilkins, never having seen Randall's letter to her containing the poisonous phrase – 'as far as the experimental X-ray effort is concerned there will be at the moment only yourself and Gosling' – expected that he, with Stokes, would analyse the photographs that Rosalind and Gosling took.

(Decades later Wilkins reasoned that Randall deliberately manipulated the misunderstanding, in order to push him aside and himself get back into what was revealing itself as the most exciting project in biophysics.) For all his building of a big department, Randall had no piece of the action himself. Others have speculated that Randall wanted to get into the work because he was disappointed with the slow progress King's had been making under Wilkins.

News of the row swiftly spread to others on the King's team who had come up to Cambridge. When they went punting on the Cam, Geoffrey and Angela Brown were with Maurice in one boat, Rosalind with friends in another. Maurice looked up and saw Rosalind with the punt pole bearing down on them in what he saw as a menacing way. 'Now she's trying to drown me!' he said. They all laughed, but it wasn't funny.

Rosalind was not one to do without a proper holiday. Hearing that Margaret Nance, her Unesco friend from Paris, now in London, was organising a party to go in August to the Ile de Batz off the coast of Brittany, she asked if she might join. Arriving after the others, Rosalind found that there was no accommodation for herself and another late arrival, Norma Sutherland. After spending one night sleeping on the floor of a barn, the two women decided to return to the mainland to seek accommodation. There appeared to be none, but Rosalind, with her determination and excellent French, found rooms. The holiday went well; a photograph taken at Quimper shows Rosalind, with a full-skirted floral print twinned with a tailored anorak, the most stylish of the trio. Norma, seeing her at close range, then, and over the next few months, got a clear view of a double-sided personality:

> Her manner was brusque and at times confrontational – she aroused quite a lot of hostility among the people she talked to, and she seemed quite insensitive to this. But she was kindness itself to me and I have fond memories of that week together. I think she needed friends away from her workplace. A couple of months later I was to leave for home [Australia] and she invited me to lunch to say goodbye and presented me with a little gift to occupy the long hours on the voyage. I was touched by this gesture which showed a much softer side of the Franklin character.

After Brittany Rosalind put her bicycle onto the train and on 16 August, after travelling all night, arrived in the Dordogne and bicycled to the village of St Leon sur Vezère, where her old St Paul's friend Anne Crawford, now Piper, and her husband Michael had rented a derelict farmhouse. She came even though warned that one Piper child had chickenpox. Anne knew that Rosalind was always good with her brood – relaxed, generous, imaginative – and they spent a good week playing with the children and visiting the caves at Lascaux. As with children, so with landscape; in both Rosalind found release from the tensions of life and laboratory. To Adrienne she wrote from the Dordogne of 'a heavenly view across the valley and no human beings within miles'.

News of the confusion about who was doing what at King's College London reached Caltech in Pasadena. So too had word of some excellent X-ray photographs of DNA. Linus Pauling had in the back of his mind that having solved one part of the cell's mysteries, the protein, he might as well get the nucleic acid as well. Brazenly he wrote and asked Randall to send to him the DNA X-ray photographs of DNA, as he understood from a friend of Wilkins that Wilkins was not planning to interpret them.

Randall gave the great man a dusty answer. 'Wilkins and others,' he replied, were 'busily engaged' in working out the interpretation of the desoxyribose nucleic acid X-ray photographs and it was natural that Wilkins should wish to carry on. He added, with an ethical flourish, 'It would not be fair to them, or to the efforts of the laboratory as a whole, to hand these over to you.' In any case, he told Pauling, as Wilkins was to speak about his work at a conference in the United States, there should be no mystery about what King's was accomplishing.

Before Wilkins left for the States and while Rosalind was away, he asked his colleague Stokes if he could work out what sort of an X-ray pattern a helical structure would give. Indeed Stokes could. It took him twenty-four hours, the crucial calculations being done on his commuter train home to Welwyn Garden City. They were correct. Stokes, like Avery, an unusually quiet and reserved man, allowed later that he sometimes thought his result 'worth perhaps 1/5000 of a Nobel prize'.

Wilkins put the information in a letter to Crick, saying that he and Stokes now felt that the DNA molecule had the shape of a helix. To make his point, he sketched a diagram of a helix in the margin. Crick's reply was that Wilkins was wasting his time; the protein molecule, not DNA, was the thing to study.

Not to Stokes. His mathematical diagram of undulating lines made such a pretty picture that he named it 'Waves at Bessel-on-Sea'. (Bessel functions are mathematical devices used in structure calculations; Bexhill-on-Sea is a seaside town in Sussex.) Wilkins pinned it up on the laboratory bulletin board and left for his Gordon Conference (part of an annual series of summer conferences on scientific topics, held at a school in the White Mountains of New Hampshire). He also left Rosalind a note.

> Stokes has supplied a few good things on helices & I have done more swelling & shrinking of fibres. The cross section increases by a factor between 3.7 & 4 on swelling to 100% humidity, Length 40–30%.
>
> I think these simple *volume* experiments probably good

> enough to tell the no. chains p.u. cell [number of chains per unit cell].

Wilkins ventured some suggestions about the density and the pitch of the chain, or chains, and concluded cheerily, 'Hope you have a good holiday, MW.'

The note was waiting when Rosalind got back from France. She had had a splendid holiday; now it was spoiled. She did not find Stokes's wave patterns pretty, but rather an intrusion on her assigned territory.

She had been getting remarkable results. The only researcher working at high humidity, determined to get the best possible photographs before embarking on interpretation, she took sharp clear pictures that revealed something no one had noticed before. There were two forms of DNA. When hydrated, the fibre became longer and thinner. When placed over a drying agent, it changed back. This transformation explained why Astbury's patterns had been difficult to interpret further. All earlier attempts to understand DNA's structure had been looking at a blur of the two forms. Rosalind and Gosling were tremendously excited by this finding.

They called the new, longer, thinner, heavily hydrated DNA, 'wet', or 'paracrystalline' or, more simply, the 'B' form. The other, shorter, drier alternative, which they could reproduce at will, was 'dry', 'crystalline' or the 'A' form.

This achievement was essential to the great discovery that lay in wait. Rosalind's skill in chemical preparation and X-ray analysis that Bernal later called 'among the most beautiful X-ray photographs of any substance ever taken' had given the first clear picture of DNA in the form in which the molecule opens up to replicate itself.

When Wilkins returned from the United States, he saw Rosalind's new sharp B pattern and was impressed by its corroboration of Stokes's Bessel calculations. He called this to her attention. Might other kinds of DNA give a B picture as well? He now had

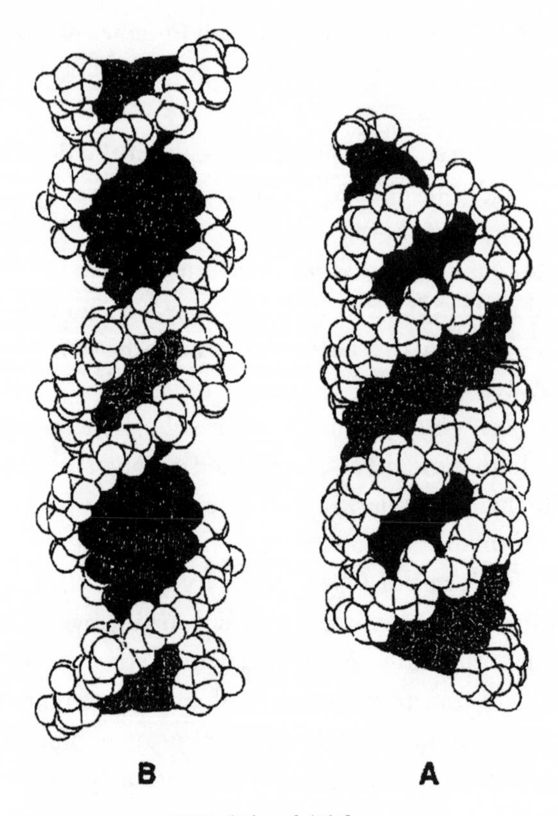

DNA 'A' and 'B' forms

some DNA samples of his own, given to him by Erwin Chargaff at Columbia. Perhaps the three of them – Franklin, Wilkins and Stokes – should collaborate?

Collaborate? She exploded. It was the only time that Wilkins saw her completely lose control. 'How dare you interpret my data for me!' Rosalind had good reason. Her work was her life, the core of her identity. Undervalued at King's, she had just achieved extraordinary results by working in virtual isolation. Now what she saw as a less able colleague of higher rank was proposing to elbow in and spoil the clarity of her investigation.

Where Wilkins saw rage, one of her brothers saw despair. One night when Colin Franklin and his wife Charlotte were expecting Rosalind to dinner at their home in Hampstead, he was startled to see her arrive in tears – 'weeping' as he recalled, 'beyond redemption'. In the Franklin manner, he pretended not to notice and said nothing.

By late October Rosalind was so miserable that she was ready to abandon her work. She poured her heart out to her old friend and confidante, Adrienne Weill.

> Dear Adrienne,
> I'm sorry I have been silent so long. I came back from my holiday to the blackest of crises in the lab, which lasted weeks and weeks, took all my energy and left me too fed up to write to anybody. Things have calmed down a little now, but I still want to get out as soon as possible, and think seriously of coming back to Paris if Paris will have me.

On 25 October 1951, with 'one more heave', the Conservatives ousted Labour in the general election and were back in power, with Winston Churchill prime minister once again. At almost the same time, Randall was forced to recognise the contest within his own borders. He called Wilkins and Rosalind to his office, for a meeting from which a treaty of sorts emerged. Rosalind, using the Signer DNA, would concentrate on the A form. Wilkins would tackle the B. From then on communications between the two virtually ceased.

Far from settling the matter, the formal division of labour made things worse, for as Rosalind took all the Signer DNA with her, plus the Phillips micro-camera, Wilkins was left with the pig thymus DNA supplied by Chargaff, and older equipment. Worse luck, the Chargaff DNA did not crystallise, presumably because of its low molecular weight, and Wilkins was thus unable to duplicate the A-B transition.

Wilkins had made a bad deal. The deal was bad for Gosling too. A PhD candidate, he found himself torn between his thesis supervisor (Rosalind) and the assistant head of his department. He was left, in another of his colourful phrases, 'acting as Envoy

Extraordinary and diplomatically carrying the "sense" of messages between them'. Freda Ticehurst felt very sorry for him: 'He got very down in spirits and worried whether he should continue working with Rosalind. But he never felt dislike for Rosalind, only sympathy.'

Why, the outsider must ask, was the Signer DNA so special? If DNA is the molecule of life and exists in every cell, why was one particular jam jar of it so prized? The answer is that the precipitate prepared by Rudolf Signer of Berne and handed out generously on a day in May 1950 had very special properties. Touched by a glass rod, the gel could be pulled out into long fibres, or strings. Wilkins had an apt comparison – 'It's just like snot!'

Fresh DNA was best obtained from animal organs. Researchers spent a lot of time at slaughterhouses and butcher's shops (where calf thymus glands were sold as sweetbreads). As there was no proper refrigeration, the organs from freshly killed animals had to be obtained before degenerative changes started in the tissues.

The next task was to extract the DNA from the organs without degrading it, then to pull out from the viscous DNA solution a single oriented fibre, position it in front of a camera and to direct an X-ray beam through it. This produced spots on the film from which to try to work out the atomic position of the repeating units in the fibre.

Signer's DNA, because of his careful and unique preparation with alcohol and salt, was superior to the rest because of its very high molecular weight. According to Paul Doty, a Harvard chemist and another lucky recipient of the manna from Berne,

> In the early 1950s little was known about preparing DNA. Most attempts led to degraded material that could not be pulled into a fibre as was needed for X-ray diffraction. Signer's preparations did produce good fibres. How he avoided substantial degradation was not clear at the time.

> His recipe was vague and clearly did not describe the whole art. Hence his samples were much desired by those who knew they gave good diffraction diagrams.

At King's, Mary Fraser, a chemist, took on the task of trying to make more DNA like Signer's, using crates of cod roe. What she produced had a molecular weight lower than Signer's but was satisfactory for making what was needed for the doctoral thesis of her husband Bruce Fraser, who was studying DNA through infrared methods. He recalled, 'We ended up smelling like a couple of Billingsgate porters. There were no showers at King's.'

After the dramatic division of labour, Rosalind threw herself into collecting and analysing her data. Her notebooks reflect intense activity: 'Either the structure is a big helix or a smaller helix consisting of several chains. The phosphates are on the outside so that phosphate–phosphate inter-helical bonds are disrupted by water.'

It was obvious to see that the phosphates were on the outside because of the ease with which the water went in and out of the fibre with changes in humidity. Phosphates soak up water – 'hydrophilic' in her terminology – causing the DNA molecule to lengthen in the process.

Wilkins found isolation less conducive to productivity. Frustrated, he began visiting Cambridge regularly to talk with his old friend Francis Crick at the Cavendish. Crick, a protein crystallographer, had met Maurice while working for the Admiralty during the war, then in 1947 when, looking for a job, he wandered into the biophysics enterprise at King's. He did not stay long. Interested in what he recognised as 'the borderline between the living and the non-living', he found the cod roe, ram sperm and calf thymus glands on the King's research menu 'rather far on the biological side of that border'.

The two men had much in common: both had been engaged in wartime weapons research, had a broken wartime marriage and first child behind them and were picking up the pieces of disrupted lives. And, yes, Crick too had been influenced by Schrödinger's *What Is Life?*

In their mid-thirties they even looked somewhat alike – very tall, lean, bony-faced. They were equally adorned with two middle initials, which looked quite grand in scientific publications: F.H.C. Crick, M.H.F. Wilkins. There was one major difference: Crick had remarried, and had a pretty French wife of artistic bent. Indeed, the Crick household in Cambridge, with a racy bohemian atmosphere, good food and scientific conversation, was a magnet to Wilkins who, German girlfriend notwithstanding, was just this side of eccentric, locked in on himself, living in a small top-floor flat overlooking Tottenham Court Road and going about in a curious long black coat with the collar turned up. Needing companionship, hot dinners and conversation, he became overly close to Crick and attracted to his wife, Odile. Her French elegance, much noticed in austere Cambridge, was summed up by another admirer as: 'she wore white dresses'.

But Crick had someone else to talk to. Jim Watson had arrived in Cambridge from Copenhagen, having wangled a transfer of his Merck–National Research Council fellowship from Copenhagen to Cambridge so that he could join Max Perutz's department to study crystallography and plant viruses. Watson told his fellowship office in Washington baldly that he now knew that X-ray crystallography was the key to genetics. A symposium that summer at his own favourite conference venue, Cold Spring Harbor on Long Island, had brought home to him even more than ever that DNA was *the* carrier of genetic inheritance. What's more, his mentor Max Delbrück had persuaded him that understanding the gene was the problem of the century. Whoever accomplished it would be covered with honour. Honour was what Watson wanted, and he headed for it in Room 103 of the Austin Wing of the Cavendish Laboratory. The small bare room contained only blackboards, tables and Francis Crick.

Affinity is no easier to explain than antipathy. The two men clicked just as much as Rosalind and Maurice repelled each other. Both were irrepressible talkers, with quick minds and complementary expertise. Watson knew biology and genetics; Crick was a physicist who had taught himself X-ray crystallography. The words poured out and continued every day over lunch at the Cavendish's local pub, The Eagle. Both laughed a great deal, even when, as was not always the case, they were discussing how genes might be put together – Crick in a loud bark, Watson in a snuffling snort that showed a lot of his gums. Crick had the merry, knowing eyes of the super-bright; Watson the horizon-scanning gaze of radar. Neither had an ounce of depression in him, while Rosalind and Maurice, in their very different ways, were prey to melancholy.

Nobody at the Cavendish was doing work on DNA. That belonged to King's. Unlike Watson, who had snared his doctorate at the age of twenty-three, Crick at thirty-five had yet to get his. He was working on his thesis on X-ray diffraction of proteins. While not specifically interested in DNA itself, he was curious about how genes copied themselves. From Wilkins he was informed about the possibilities of a helical structure for DNA. Watson goaded him to follow Pauling's lead and build a model. There was no need to spend any money or touch a fibre of the snot-like gel – just to make an educated guess of how the molecule might be put together according to the rules of chemistry. Then, with balls and wires from the machine shop, assemble a structure illustrating the guess in three dimensions.

Both men had been deeply impressed by Pauling's achievement, as well as by the failure of their Cavendish colleagues to discover it themselves. They were impressed too by the method Pauling had used to succeed: model-building. However, Crick was on a precipice at the Cavendish. Bragg found him irritating, intrusive, over-confident and under-productive, and hoped he would finish his thesis and be gone. There was no way that Bragg would accept Crick's diversion from his main task to speculations about DNA.

For Rosalind, Wilkins was not the only irritant at King's. He had a sarcastic office mate, Bill Seeds, a rotund Irishman with Swiss connections on his mother's side, who was a graduate of Trinity College Dublin and engaged on the mechanical details of a reflecting microscope. Seeds was the Biophysics Unit's joker, given to raucous remarks. He also loved running round the department, stirring things up, and coining nicknames: the quiet thinker Stokes was the 'Archangel Gabriel'; the lean and taciturn Maurice was 'Uncle'; Honor Fell was 'Aunty'. It did not take long for Seeds to turn Rosalind into 'Rosy'. She could not stand him. He did not like her either.

She knew very well what the common diminutive of her name was – she had an Aunt Rosie – but she herself had long succeeded in being called 'Ros-lind', pronounced in two clipped syllables. From her family and close friends, she would accept 'Ros'.

It might have helped if Rosalind had recognised that all at King's had nicknames. To Dorothy Hodgkin's biographer, Georgina Ferry, these had a double function: to demystify superiors but also to get round the stilted custom then prevailing in British offices, laboratories and public schools in which males addressed even their best friends by surname alone, while a woman's surname was always prefaced by 'Miss' or 'Mrs'.*

Rosalind and Seeds had a run-in over the design for her tilting camera stand. Seeds felt the design was poor; he did not want scarce workshop time wasted, so suggested some changes. Rosalind refused to discuss them. What he saw as her 'That's the way

* This diminutive is not as inherently insulting as many critics of Watson's 'Rosy' in *The Double Helix* have suggested. When I first arrived in England from Massachusetts, I shared a flat with an Anglo-Irish Reuters journalist named Rosanna Groarke, and routinely addressed her, although no one else did, as 'Rosie'. It seemed a sign of American friendliness.

we did it in France and that's the way that it's going to be done here' attitude brought out the worst in him. Once when she had covered some apparatus in the lab with blackout cloth, she came back to find the sign 'Rosy's Parlour' hung on it. This hint of the gypsy, the alien and the occult shows the swart associations that Rosalind stirred at King's.

The prank sent her into another tantrum and she called them all 'little schoolboys'.

Early in November, Crick invited Wilkins to visit him at his home in Cambridge for the weekend and Wilkins talked freely. Watson was there too. They pumped Wilkins for information and he repeated much of what he had said at the Perutz meeting in July, interspersed with comments on what had become an obsession, his difficulties with Rosalind. She had, he felt, virtually shut him out of his own subject. She had been given the best cameras and best DNA, only to keep her results to herself instead of sharing them with her group in the spirit of scientific collegiality.

Wilkins had no reason not to talk freely. As far as he knew, nobody at the Cavendish was working on DNA. Watson and Crick kept urging him to follow Pauling's example and build models. But Rosalind would not hear of it. There was no way to prove that any model – no matter how beguiling – reflected reality. Had she not heard Bernal say in Stockholm that any proposed molecular structure should be tested against the X-ray evidence? Such evidence was what she had been enlisted to find. Rods, metal plate, wires, coloured plastic balls, to her were pointless diversions, signifying nothing. The time to build models, she told Wilkins, was when you knew what the structure was.

Not everybody at King's was opposed to model-building. In early November, Bruce Fraser, working for his doctorate with Dr William Price in the spectroscopy group (allied to the Biophysics Unit), decided to put together a model to summarise the latest DNA thinking. Fraser, who sported a black handlebar moustache testifying to his war service as a pilot for the RAF, was neutral in the Franklin-Wilkins feud. Such contact as he and his wife Mary had with Rosalind was entirely friendly. Rosalind discussed the model with him just as she had read Sven Furberg's thesis on which it was based. When Fraser asked her to guess how many chains there might be in the molecule, she ventured three. Wilkins said the same, for the same reasons: the measurements of density and water content of DNA suggested more than one chain and probably three. Two would not be enough to fill the space. (As Linus Pauling later came up with a three-chain answer, the reasoning was sound, if erroneous.)

Fraser's model of DNA, completed very quickly, was a simple structure that had what would turn out to be all main features correct except for the number of chains. It had a helical shape, phosphates on the outside, and bases stacked like a pile of pennies, separated by the 3.4 Å distance worked out by Astbury. Rosalind saw it, and her view was what she felt about all models: 'That's very nice – how are you going to prove it is the solution?'

A big step in the right direction, Fraser's November 1951 model was another glaring example of King's' institutional hesitancy. Its details were never published, just as Stokes's calculations on helical diffraction were never published. Wilkins himself did not think it worth pursuing because, although Fraser's model was full of potential, there was no structural hypothesis behind it; it did not explain anything. In a few months Bruce and Mary Fraser, with their new baby, prepared to emigrate to Australia – out of the drama until its very end.

On 21 November King's held a colloquium on nucleic acid structure. There were three speakers, two of whom were barely speaking to each other. First, Wilkins repeated what he said in Cambridge in July. Next Stokes summed up his work on helical theory. Finally, Rosalind presented her findings about the A-to-B transition in hydrated DNA. The notes she made to prepare for the talk show much mention of the helical possibilities of DNA, but in the memories of those present, all admittedly vague, she did not utter the critical word 'helix' to which Wilkins and Stokes had just given so much attention.

None of the three speakers suggested any connection between form and function – that is, they did not anticipate that seeing the shape of the genetic molecule would reveal how it did its work.

If Rosalind could have known that her entrance on the world stage would be dominated by what she wore on 21 November 1951, would she have dressed differently? Almost certainly not. For her presentation she wore what she felt was appropriate for work: probably a white blouse, dark skirt and lab coat. Watching her intently from the small audience of fifteen or so was Jim Watson. He had asked Maurice if he might attend. He hoped to learn whether Rosalind's new X-ray pictures might yield any support for a helical DNA. Also, having heard about the terrible Rosy from Maurice, he was keen to get his first look.

Girls, for Watson, were equal to genes as objects of supreme curiosity. He appraised her against his standard of feminine beauty – Hedy Lamarr, the French 'popsies' studying English at Cambridge language schools, and his sister Elizabeth – and found her wanting. As he recorded in *The Double Helix,* in a sentence that has since become notorious: 'Momentarily I wondered how she would look if she took off her glasses and did something novel with her hair.'

Believing that young women should add sparkle to life, Watson did not think much of Rosalind's dress sense either:

> Though her features were strong, she was not unattractive and might have been quite stunning had she taken even a mild interest in clothes. This she did not. There was never lipstick to contrast with her straight black hair, while [he added ungrammatically] at the age of thirty-one her dresses showed all the imagination of English blue-stocking adolescents.

Watson had no idea of Rosalind's scientific reputation, let alone her background. An acute observer of physical detail but a poor psychologist, he assessed her as the 'product of a mother whose own career aspirations had been thwarted' and who 'duly stressed the desirability of professional careers that could save bright girls from marriages to dull men'.

The very opposite was true. Rosalind had made herself into the antithesis of her mother. The last thing that Muriel Franklin wanted, for herself or her daughter, was a career, and, being married to a man she found the most fascinating in the world, she was incapable of imagining the plight of a woman chained to a bore.

Thus preoccupied, Watson did not grasp what Rosalind said about the water cloaking the phosphates on the outside of the molecule. He took no notes. He never did.

One week later he and Crick had put together a model of their own. Using leftover parts from a previous Cavendish attempt at a different exercise and – not to be overlooked – their long hours of theorising about helical diffraction and folding polypeptide chains, they assembled a structure based on what Watson had half-gleaned from King's and from what Maurice had told Francis from memory earlier that month. Their operating principle was clear, and it was the opposite of Rosalind's: to incorporate the minimum number of experimental facts.

Their model was largely posited on their own understanding of the stereochemical requirements of the model and on Watson's memory of King's research data. That their immediate inspiration was Rosalind's talk was made perfectly clear in a written summary of their approach to the model: 'Stimulated by the results presented by the workers at King's College, London, at a colloquium given on 21st November, 1951, we have attempted to see if we can find any general principles on which the structure of DNA might be based.' Their handiwork resulted in a three-chain model of a helix with the phosphates on the inside, bases on the outside, sticking out. John Kendrew insisted that out of courtesy, the

King's team – Wilkins, Franklin, Seeds, Gosling and Fraser – should be invited up to see what the Cavendish pair had done.

The gross miscalculation was obvious. Rosalind wasted no time on pleasantries. Where was the water? She pointed out that DNA is a thirsty molecule – soaking up water more than ten times what they had allowed. The phosphates had to be on the outside, encased in a shell of water. If the molecule were as Watson and Crick suggested, how could it hold together? The sodium ions that they had erroneously positioned on the outside would be encased in sheaths of water and thus unavailable for binding.

Rosalind was polite enough as she spoke – according to Seeds (although Watson later described her as 'pugnaciously assertive'). On the train home, however, with Wilkins and Seeds, she was jubilant. She had not expected the model to be right. The whole approach was unprofessional. The way to proceed was not to make a hypothesis until the experimental facts were in hand, then not to publish any results until the facts were absolutely certain. However, Gosling surmised she had learned something and that now she was a little bit worried about how much one ought to talk about what one was doing.

Crick later took the blame on himself. The model was completely wrong because 'in fact there was a lot of water there. I did not know enough chemistry to know that things like sodium are highly likely to be hydrated anyway.'

As a result of their fiasco, Bragg ordered Watson and Crick to a halt. He and Randall, after a quiet word, agreed that investigating the structure of DNA would be left to King's. (This pact, to various American accounts, manifested a very English sense of gentlemanly fair play. This interpretation ignores the fact that in post-war Britain queuing and sharing had become part of the national genius and that, with both laboratories supported by the same funding council, the MRC, it seemed to both sides almost criminally wasteful to duplicate research.) In any event, as a testament to their assent to the agreement, Crick and Watson sent their jigs – their bits of metal and wire – down to King's.

Every Christmas before the college closed for the holiday, King's Physics and Biophysics Unit held its annual dinner. This was another of Randall's command performances. Its highlight was a rollicking pantomime at which few were spared, least of all 'JT' himself. By all accounts, the 1951 entertainment was unusually lively, so much so that the devout Professor Coulson complained to Randall that the highjinks had overstepped the mark. Some women wore balloons inside their blouses. Condoms were inflated and released in the Strand. Music was performed on a piano hijacked from the theology department.

The central sketch portrayed a PhD candidate, played by Derek Mould, undergoing his viva examination. Bill Seeds was the external examiner just arrived from Ireland by donkey, and Stan Bayley was Randall, with bald-pate wig, glasses, pipe, linen jacket and bow tie. On the desk was the tatty rubber plant that was Randall's pride and joy.

The scientist-at-play lyric is a demanding form. Abstruse technical terms must be woven with gibes at laboratory personalities in verses with a strong beat and a repetitive chorus. The creative brains of King's Physics and Biophysics were well up to the challenge:

> Alex Stokes had an equation – For the tracks
> at every station,
> And the lines at Clapham Junction – Were a
> complex Bessel function!
> Now the boss's name is Randall – He's the guy
> who turns the handle,
> And we sin and slave and MURDER – Just to
> spread his empire further!

Chorus:
Have you ever seen –
Have you ever seen,
Such a funny lab before.

Rosalind had never seen such a funny lab before, and she was determined to get out of this one as soon as possible. Over the holiday break she set out for Paris to try to get her old job back.

ELEVEN

The Undeclared Race
(January 1952–January 1953)

BACK IN HER BELOVED PARIS, Rosalind brought her photographs with their many reflections to show Vittorio Luzzati. He suggested that she apply a technique called Patterson function analysis to interpret them. 'It's what a crystallographer would do,' he told her, seeing her as a physical chemist seeking guidance.

Jacques Mering, however, was unpleasant. She had left his lab and that was that. Once more he refused to lend his name to the papers she was writing based on the results they had achieved together. Adrienne Weill was well aware of Rosalind's feelings for Mering, which she saw as deep, genuine and somewhat tragic.

Back in London empty-handed, Rosalind took a bold step. She went to see J.D. Bernal at Birkbeck College and asked to join his biological group. To her relief, the great man welcomed the idea – in principle. She dared not press further to ask if she might move to Birkbeck that very year. There was no alternative but to return to King's College London, concealing that defection was in her mind.

She announced to Gosling that they would apply the Patterson procedures to measure the reflections of DNA's A form that they would get once their designed tilting camera was ready. It would allow them to photograph the fibre from a great many angles. They were following Luzzati's advice, to be sure, but it was advice Rosalind was very willing to take, for it was what she would have wanted to do anyway. The deductive approach was what Bernal had recommended in Stockholm (where she had noted 'Real test requires scattering out in crystal diagrams and final structure by Patterson sections'). There was nothing wrong about the method except its difficulty and inherent limitations. Crystallographers long after the event agreed it was perfectly reasonable for Rosalind and Gosling to proceed in this manner.

'The Patterson' was named after A. Lindo Patterson, a New Zealander educated at McGill University in Montreal, who in the 1920s had developed the method for circumventing the difficult 'phase' problem of measuring X-rays' peaks and troughs. It relies only on the intensities – the blackness – of the spots captured on a photographic plate. A Patterson map, which looks like a contour map, is a complicated overlay of mathematical pictures showing the heavier atoms, such as phosphorus, standing out as peaks. It makes it possible to estimate the distances between these atoms. By methodically registering the intensities (or blackness) of the spots in the X-ray diffraction picture, with luck, glimpses of the internal configuration of the molecule may slowly emerge. Carolyn Cohen, a protein crystallographer, thinks of it as 'the structure speaking to you through those spots'. The distinguished crystallographer Dorothy Hodgkin of Oxford, Bernal's collaborator, a Fellow of the Royal Society since 1947, was considered 'the queen of the Patterson function'. David Sayre, who had worked with Hodgkin, said, 'That was her tool and she was unbelievable at it . . . She thought in terms of maps.'

The work suited Rosalind's mathematical brain. It required intense concentration during long hours of laborious calculation – what is now done in a flash by computer. She and Gosling, having determined the intensities of the spots, then entered into a calculating machine the numerical equivalent of the sequence in which the spots should occur, obtained from numbered strips of cardboard named after their inventors, Arnold Beevers and Henry Lipson. Gosling had nightmares that the handsome mahogany box of sequentially numbered Beevers–Lipson strips would drop to the floor and they would have to reassemble them.

Rosalind felt that if she and Gosling measured all the vectors (directions) between the atoms in the various diffraction patterns they obtained from the A form of DNA, the data would tell them about the three-dimensional configuration of the molecule. 'That is what she felt she was there to do, rather than speculative model building,' Gosling later explained.

Rosalind had not done Patterson calculations before. What's more, no one had ever attempted 'cylindrical sections' – an order of magnitude more difficult, like trying to do a three-dimensional jigsaw puzzle. The difficulty did not daunt her.

Her other task early in 1952 was to write an account of her previous year's work for the Turner and Newall Fellowship. In a detailed report, dated 7 February, she said she had spent the first eight months assembling the necessary apparatus and then moved on to undertake 'a systematic X-ray investigation of the fine fibres of desoxyribose nucleic acid obtained by M.H.F. Wilkins from the specimen prepared by Professor Signer (Berne)'. She then gave the dimensions of the unit cell of the molecule. (The unit cell is to the crystal what a brick is to a wall: the basic, repeating, building block.) She indexed the unit cell tentatively, according to the *International Tables of Crystallography* as 'face-centred monoclinic'. The DNA molecule, she said, had its phosphates on the outside, and changed from the crystalline to the paracrystalline (that is, the 'wet', or gelatinous) state upon hydration, with a corresponding change in the length of the fibres. The results, in sum, suggested 'a helical structure (which must be very closely packed) containing probably 2, 3 or 4 co-axial nucleic acid chains per helical unit, with the phosphate groups near the outside'.

The report was a succinct summary of an impressive year's work.

Other events in February 1952 included the death of King George VI in his sleep after a long illness widely reported abroad but hushed up in the British press, and the refusal by the US State Department of a passport to Linus Pauling. Under the McCarthy era's McCarran Act, travel abroad was forbidden for Americans with suspect political sympathies. Pauling's active opposition to the Cold War, nuclear weapons and the development of the hydrogen bomb was a prime example. Of narrower interest was *Acta Crystallographica*'s receipt on 16 February of a paper by W. Cochran, F.H.C. Crick, and V. Vand offering a mathematical prediction of the pattern that atoms arranged in a helix would form when diffracted by X-rays. The authors acknowledged that the same theory 'was also derived independently and simultaneously by Dr. A.R. Stokes' – information they attributed to 'private communication'. (Stokes, still reluctant to publish his own helical work done at King's, declined to be cited more directly in their paper.) Helices were indeed, as Crick had commented about this stage, 'in the air'.

> What is it that makes one's home country seem so awful after returning from living in a foreign country one has come to love?

Throwing this bitter question to Anne Sayre, Rosalind showed clearly that part of her aversion to King's was an aversion to home. She spelled it out in a letter of devastating candour to the Sayres:

> What you feel about America is exactly what I feel about England . . . To put it at its lowest, I suspect that I enjoyed being a more interesting person in France than I am in England – more interesting simply because English scientists are rarer in France than in England.

At King's, Rosalind acknowledged, her equipment and facilities were good – 'in fact scandalously good, considering the shortage of money generally for such things'. But (as she had told Adrienne Weill the previous October), 'I still want to get out as soon as

I can.' There followed some very harsh words about her colleagues:

> The very young are mostly thoroughly nice but none of them brilliant. There are one or two more serious people who are good and pleasant, but refrain from doing research so as to be able to keep outside the unpleasant atmosphere. And the other middle and senior people are positively repulsive, and it's they who set the general tone. I've got myself organised so that I hardly ever see any of them, which makes things better but distinctly boring. The other serious trouble is that there isn't a first-class or even a good brain among them – in fact nobody with whom I particularly want to discuss anything, scientific or otherwise, and I so much prefer to work under somebody who commands my respect and can offer some encouragement.

To the Sayres, she confided the news that Bernal had welcomed her application to join his department. Birkbeck would suit her better, she thought; as an evening college, unlike the rest within the University of London, it took only part-time students who must be people who really wanted to learn. Then, in a rare and explicit reference to her Jewishness, Rosalind added:

> And they seem to collect a large proportion of foreigners on the staff, which is a good sign. King's has neither foreigners nor Jews.

Rosalind had a tendency, as her sister Jenifer recognised, of 'retreating into a disconcerting silence or obvious disapproval if she felt out of tune with her surroundings'. At King's Rosalind felt out of tune to such an extent that she misinterpreted the facts. Simon Altmann and Louise Heller, to name but two, were both Jewish and foreign. Attempting to list others would be invidious. Jack Lowy, a Czech; Stephen Pelc, the Austrian Marianne Friedlander? No one was counting, or even speculating as Rosalind clearly had done, whether a surname indicated Jewishness. There were many at King's who did not realise that she was Jewish. However, King's seems to have touched that nerve in her alert to even a hint of anti-semitism. In Paris she had been surrounded by people generally known to be Jewish; anybody who had survived the Nazi occcupation was not inclined to keep the fact quiet. At King's when Rosalind did not see Jews in the significant figures around her – Randall, Wilkins, Coulson – she wrongly registered them as totally absent. In many ways, her demeanour at King's fits the description of the twelfth-century English Jews of Scott's *Ivanhoe:* 'watchful, suspicious and timid – yet obstinate, uncomplying, and skilful in evading the dangers to which they were exposed'.

Why not come to the United States? Anne Sayre countered. David, her husband, could organise finance for a period of work, and the eponymous Patterson was in Philadelphia (at the Fox Chase Cancer Center), with 'stacks' of equipment which he would lend cheerfully. 'My own anti-Bernal feelings', Anne confessed, 'are passionate', and she warned Rosalind that his staff rarely saw him. (Bernal, a zealous Communist, travelled widely and was heavily involved in international peace and Eastern European scientific organisations.) Whatever Rosalind decided, her good friend concluded, 'if you murder anyone at King's, I will fly over as a character witness and swear it was justifiable homicide'.

Bernal's politics did not concern Rosalind at all. 'Whatever one may have against the man, he's *brilliant,*' she said. That was the kind of man she wanted to work with.

Rosalind's unhappiness at King's left a legacy of sour memories of an unhappy young woman going about with a 'face like a thundercloud' and who at best looked sad. They speculated on the cause of her unhappiness. Was it some emotional attachment in Paris that had gone wrong? At worst, she grew angry and displayed the rage the French shopkeeper must have seen: eyes blazing, words pouring out. Geoffrey Brown, a research student in the physics department, had an altercation with Rosalind over a Tesla coil (for detecting leaks in vacuum systems). She asked to borrow the one he had been allocated by the physics department. When she did not return it upon his request, 'I went and took it back, and screwed it onto the wall. And she came in, pulled it off and walked straight out.' He did not try again. She was senior to a mere graduate student.

Another doctoral student, John Cadogan (later Sir John), recalled working late in the physics laboratory one Saturday night; it was the only time he could get at the infrared spectrometer; his own chemistry department didn't have one. Suddenly the door opened and he found himself facing a livid Rosalind: 'She nearly scared the living daylights out of me. She demanded to know what I was doing there at that time of night. She threatened to call security.' When Cadogan identified himself, she calmed down, whereupon a familiar pattern asserted itself: 'She became very kind and warmed to me, interested in the work of a young researcher and explained to me much over the coming months about the progress of the DNA work.'

There is little doubt that these ugly incidents occurred, nor that the male consensus in the lab was that most objectionable of male compliments: 'She looked beautiful when she was angry.' All such anecdotes are one-sided and tainted by hindsight. Unfortunately, Rosalind's side of these encounters was never recorded.

Maurice Wilkins bore the brunt of her antagonism but felt he was getting used to it. He wrote to Francis Crick in the spring of 1952, 'Franklin barks often but doesn't succeed in biting me. Since I reorganised my time so that I can concentrate on the job, she no longer gets under my skin. I was in a bad way about it all when I last saw you.'

When Rosalind told Anne Sayre about the 'one or two more serious people who are good and pleasant but refrain from doing research so as to be able to keep outside the unpleasant atmosphere', she clearly meant Alec Stokes. Stokes, in fact, much preferred theory to fumbling with cameras and messy gels. In any event, he got on well with Rosalind and invited her in 1952 to give five lectures on crystallography. Her thick sheaf of notes shows that she took a lot of trouble to explain the basics of X-ray diffraction to honours students studying for the bachelor of science degree.

That spring saw the successful completion of the tilting camera that she and Gosling had designed. All labs at that time were very reliant on their workshops, as so much had to be made by hand. King's prided itself on an excellent shop, which had managed to fashion parts out of army surplus equipment 'liberated' from German laboratories, such as a damaged Siemens electron microscope. A King's technician was startled one night to see Rosalind whirl in, in evening dress, almost unrecognisable from her daytime persona. A swift adjustment to some detail and she was gone. He felt he had been visited by a creature from another planet.

With her photographs increasingly clear and sharp, Rosalind took a set to Oxford to show Dorothy Hodgkin. They spread them out on the big tables of the shabby museum laboratory where Dorothy worked. ('Like most British labs,' an American visitor commented, 'it looks like the corner of a dusty old barn.') Hodgkin exclaimed that Rosalind's photographs were the best she had ever seen – so clear, in fact, that it might be possible to work out the space group of the crystal. (According to the *International X-ray Tables,* crystals are classified by their symmetry – that is, the shape of their unit cell – into 230 space groups. Prepared by W.T. Astbury of Leeds, with Kathleen Lonsdale, at the Royal Institution and published in 1924, the tables were, in Rosalind's time, the crystallographer's Bible.)

Rosalind volunteered that she had already narrowed the possibilities down to three space groups. But two of these, Hodgkin swiftly pointed out, were impossible. Only one had the correct 'handedness' for the sugars in DNA. Rosalind accepted the correction.*

* Watson has often used this story against Rosalind, claiming for instance in his Harvard lecture of 30 September 1999, that she was stung by Hodgkin's comment which she took as a rebuke for her ignorance of crystallography. But Jack Dunitz who was present, says that Watson's retelling of Rosalind's reaction is exaggerated and that she accepted the correction of such a respected senior scientist with good humour.

Rosalind never doubted that the B form of DNA was a helix. How could she have done? That was the general view at King's. It was what she said in her Turner and Newall report and it was what Randall reported to a meeting of the biophysics committee of the Medical Research Council on 14 March 1952. The MRC committee, formed in 1947, oversaw all biophysics research expenditure and met regularly at its recipient institutions.

That day the MRC committee gathered in London at the Department of Clinical Research of University College Hospital Medical School. Randall's progress report was one of two items on the agenda. At King's, he announced, the apparatus-building phase was over; now they were conducting a concentrated attack on cell structure. In sum, 'From the X-ray evidence nucleic acid was now thought to have a helical structure.' Randall invited the committee to hold its next meeting at King's, where the new physics department was nearly finished. The visit was arranged for December.

For her Easter break in 1952 Rosalind went on yet another walking tour of North Wales. Her companion that year was Margaret Nance, who had accompanied her to the Italian Alps and who often, now that they were both back in London, joined Rosalind in long walks around London, to Windsor Great Park or the Berkshire Downs. Rosalind would ring up, and, according to Margaret, 'would not take no for an answer. If I refused, she wanted to know why.' But Margaret liked her, in spite of her occasional fierceness, for her vigour and her great love of the outdoors.

On 1 May 1952 the Royal Society held a meeting in London on proteins. Pauling, who was to have been the guest of honour, was absent. There was general astonishment that (even though the excesses of McCarthyism were well known) such an eminent scientist should be banned from leaving his own country. During the day Rosalind was approached by Pauline Cowan, a student of Dorothy Hodgkin's, who was wondering whether to accept an offer from King's. 'Don't' was Rosalind's advice. King's was not a happy place, she said, and contrasted badly with Hodgkin's cheerful and productive lab in Oxford.

Jim Watson was also there. He was still at the Cavendish in Cambridge, his work now supported by the National Foundation for Infantile Paralysis. Thwarted in his DNA pursuit by Bragg's ban, he was working on ribonucleic acid, RNA, in the form found in the virus that mottles tobacco leaves: tobacco mosaic virus, or TMV.

Watson had not forgotten about DNA, however, and wrote to his mentor Max Delbrück that Maurice Wilkins at King's had obtained 'extremely excellent' X-ray diffraction photographs of DNA, adding:

> It is obvious that a great deal of work should go into elucidating this structure. However, the people at King's are involved in a fight among themselves and so at present no real effort is being made to solve the structure. We have attempted to interest them in attempting to build plausible models. However, we have temporarily stopped for the political reason of not working on the problem of a close friend. If, however, the King's people persist in doing nothing, we shall again try our luck.

After the Royal Society meeting, Rosalind showed some of her DNA photographs to Robert Corey, Pauling's colleague and an eminent protein chemist, who had attended in Pauling's place. Corey carried the news back to Caltech. Rosalind's photographs were superb. Pauling himself had taken a few unsatisfactory photographs of DNA that year and his interest was whetted by what Corey told him.

While she was at the Royal Society, her camera was on. The X-ray photographs she and Gosling were taking required long exposure – as much as 100 hours – of a single DNA fibre positioned at very close range – 15 millimetres. Sometimes during exposure, the fibre would change from the crystalline A form to the paracrystalline B form. Once this happened so abruptly that the fibre fell off the holder. The photograph taken between 1 and 2 May showed a stark X, formed of tigerish black stripes radiating out from the centre. The spaces between the arms of the X were completely blank. It was the clearest picture ever taken of the B form of DNA, unquestionably a helix. Rosalind numbered it Photograph 51 and put it aside, to return (as agreed with Randall) to the puzzle of the A form.

This decision, in the light of subsequent events, has been much criticised. But she had every reason to continue interpreting her A-form patterns, for these contained more of the information useful to crystallographers. She refused to see a helix in the A form without more proof. She had two serious papers in preparation announcing her discoveries and preferred to wait until she was sure. What was the rush?

Struck by the macroscopic (that is, visible to the naked eye) change in the fibres between the A and the B forms, she wondered whether the molecular structure could have been changed and that the helix of B could have come unwound. Her dilemma in July 1952 is described by Robert Olby in the *Dictionary of Scientific Biography:*

> the cylindrical Patterson function obtained by Gosling strengthened her antihelical views, although the arrangement of peaks was consistent with a helix. She was misled, by what appeared to be clear evidence of a structural repeat at half the height of the unit cell, into ruling out helices for the A form, since no DNA chain could possibly be folded into a helix with a pitch equal to half the height of the unit cell (14 Å).

Everything in her education and background had taught her to be absolutely sure of her facts before she presented them to the world. No less than Dorothy Hodgkin in retrospect agreed with the soundness of Rosalind's approach. 'As Rosalind was necessarily involved in collecting the accurate data on DNA, it was natural of her to postpone model building until her data was complete, and until she had extracted all the information she could that would limit the kind of model she should build.'

For Rosalind the price of sticking to her guns was condescension. At a meeting at the Zoological Museum in Cambridge she ran into Francis Crick in a queue and told him about the asymmetries she was getting in the A form. Crick warned her that she was being misled into thinking that the A form was not a helix; that there were other explanations for what she was seeing which would accommodate a helical structure.

Crick had spoken condescendingly. Later he was to say, 'I'm afraid we always used to adopt – let's say a *patronising* attitude towards her. When she told us DNA couldn't be a helix, we said, Nonsense. And when she said but her measurements showed that it couldn't, we said, "Well, they're wrong." You see, that was our sort of attitude.' Crick, in this case, was right: the kind of asymmetry that Franklin was finding in the X-ray reflections could be accounted for. But she too was right in her defensiveness; she knew when she was not being taken seriously.

If she had felt very confident and supported, she might have been able to make outrageous leaps of imagination. 'Lack of intuition' does not begin to cover the lattice of emotions, warnings and sensitivity to a hostile environment that told her to be wary and proceed with caution.

In May, her reputation in coal research brought her an invitation from an unexpected quarter, Yugoslavia. She went off happily, having heard from Bernal that he would take her at Birkbeck any time that Randall agreed. She thought it politic to postpone telling Randall that she was leaving until she got back from her month's travel.

To be abroad once more was bliss – apart from the large cities: 'after 2 days in Zagreb I shall be able to spot a Croat a mile off for the rest of my life'. As for Belgrade: 'a pathetic place, where nothing is permanent and nobody knows whether they belong to east or west'. Otherwise, she encountered impressive research, kind people and spectacular scenery, marred on occasion by political proselytising.

In Zagreb, where she went to give a lecture on 'Some aspects of the ultra fine structure of coals and cokes', arranged by two university professors, she met Katarina Kranjc, the first woman in Yugoslavia to use X-rays and the first to take a PhD in physics. Only five years older than Rosalind, Kranjc went to the station with a large bunch of red roses. 'I do not know what I imagined a woman scientist should look like,' she wrote later, 'but I certainly did not expect her to be such a charming young girl: my astonishment was enormous.' Kranjc observed also that Rosalind was very concerned with her health and always peeled peaches before eating them.

The lecture was a great success. Visits by foreign scientists were rare and Rosalind's talk was their first direct information about crystallographic laboratories in Britain and France. Afterwards Kranjc and another chemist working on X-ray structure analysis, Dr. Drago Grdenič, took Rosalind to the opera and a theatre café. Returning to her hotel, she gave them small packages of tea and coffee as presents. They all spoke French together and her new acquaintances thought they had never heard any English person speak it better.

The trip included two days' walking and a night in a hut in the Julian Alps, and a week, which she took as holiday, on the Dalmatian coast. Dubrovnik was lovely; at Split the local scientific institute took her on a boat trip to a nearby island, with experts to expound on everything from microbiology to the economics of fishing. 'They reckon that they have already succeeded in increasing the fertility of the sea 40-fold in certain regions.' At Kricula, 'the loveliest of all', she enjoyed the company of a young Italian schoolmaster who had been asked to show her around:

> we wandered round a series of deep blue coves amid rich tropical vegetation, and bathed in the sea. When I told him that in England if the sea was more than a few metres deep

> you couldn't see the bottom he thought it the greatest joke. He said he really must go there one day, just to see – which I thought was very broad-minded of him. I'm sure if I lived in Kricula I should have no urge to travel. Then we talked politics, and he produced the most preposterous communist propaganda statements about England, which my Italian was utterly unable to answer adequately.

After passing through Venice, then stopping for two days in Paris, she was home. It was, she wrote the Sayres, 'all too good to be true – I imagine I shall have to wait a long time for such a thing to happen again'.

In June 1952 the hole in King's courtyard was filled at last. The new Physics and Engineering laboratories were opened, dedicated on 27 June by Lord Cherwell, Churchill's scientific adviser. For the occasion, the Wheatstone Physics Laboratory issued a pamphlet relating its long history, plus an account of the current research activities of the department. These included the work of the group on 'Molecular Structure of Nucleic Acids and Nucleoprotein in Cells', whose members were listed as 'M.F.H. Wilkins, R.E. Franklin, A.R. Stokes and R.G. Gosling' and their work summarised:

> The pure sodium salt of desoxyribose nucleic acid may be extracted from cells and crystallised, and X-ray diffraction shows the crystal to consist of a parallel arrangement of helical polymer chains. A very similar arrangement of chains also exists in nucleoproteins in living cells.
>
> The research, it was hoped, might cast light on the biological function of nucleic acid and even on specific molecular structures which may be associated with gene action.

King's got little credit for getting this far, because it did not go on to discover what linked the parallel chains or indeed to offer any clue to 'gene action' – how the DNA molecule replicated.

* * *

While Rosalind was in Yugoslavia, Watson and Crick in Cambridge met Erwin Chargaff from Columbia University in New York City, whom they knew to be one of the world's experts on DNA. It was Chargaff who in 1950 had discovered that in DNA, the number of adenine molecules is always nearly the same as the number of thymine molecules, and a similar correspondence is found with cytosine and guanine. In other words, there are equal numbers of purine and pyrimidine molecules. Chargaff analysed the proportions of the four bases of DNA and found a curious correspondence. The numbers of molecules present in the two bases called purines were always equal to the total amount of thymine and cytosine, the other two bases, called pyrimidines.

At Cambridge Watson was an object of curiosity for his untidiness. He went about in tennis shoes with untied laces, wore shorts even in the winter and stared at people with his mouth open, snorting with laughter. Yet all the passion Rosalind felt for France, Watson felt for England: the colleges seen from the Backs of the River Cam were for him 'the most beautiful buildings in the world'.

He had instantly judged Cambridge as a whole 'the most attractive site for science in Britain, if not the world'. Thwarted in his DNA pursuit by Sir Lawrence Bragg's ban, and determined to remain in Cambridge, he was applying himself to the related nucleic acid, RNA. He expected the RNA molecule to be in the shape of a helix – that is, a spiral, like Pauling's alpha helix of protein.

Watson brought out the motherly instincts in the Cambridge wives who saw that he was very interested in girls, but unsuccessful. Crick, on the other hand, was full of easy charm and confidence – too much for Sir Lawrence Bragg if not for the impressionable Watson.

The erudite, sarcastic Viennese Chargaff listened to the two burble on about the pitch and spacing of a likely helical structure for DNA and cast his withering eye on the pair – Watson was too gauche, Crick too slick. When he learned that Crick did not even know the chemical differences between the four bases of DNA, he formed the thought that they were 'two pitchmen in search of a helix': sharp salesmen flogging a dubious product. But Chargaff was not to have the last laugh. Even he did not know the meaning of the 'Chargaff ratios' – the strange complementary pairings that he had discovered.

Sometime between 8 June and 1 July 1952, Rosalind told Randall she was leaving King's. Did she jump or was she pushed? There is evidence both ways. Some observers maintain that Randall told her she had to go; the situation in his lab was impossible. Her own letters sound as if the initiative were entirely hers. However, Randall, in his devious way, may well have had an encouraging word with Bernal in advance of Rosalind's initial approach. What is certain is that Randall did nothing to persuade her to stay. He must have been delighted to have this easy solution to a nagging problem. The paperwork was swiftly done. On 1 July the head of her Turner and Newall Fellowships Committee formally notified Randall that Miss R.E. Franklin had applied for permission to spend the third year of her fellowship in the crystallographic laboratory of Professor Bernal at Birkbeck to 'continue to study the structure of Biological substances by means of X-ray diffraction'. Two days later Randall recommended that the move be made as from 1 January 1953. And on 21 July the chairman of the Fellowships Committee gave his formal assent.

By then Rosalind and Gosling were in a new room at King's, upstairs, more removed from the rest than before. The Patterson functions were giving what she called 'banana-shaped peaks'. Willy Seeds looking in the door got the impression of Rosalind and Gosling working on the Patterson, week after week, buried in paper, getting nowhere, buried in tables, slide rules and desk calculations requiring extreme concentration.

'We spent *ages,*' said Gosling. 'We had to think in three dimensions.' Thus months were wasted? 'Not wasted at all!' said

Gosling. As Rosalind reasoned: 'How will a model show which structure is right? The Patterson will *tell* us the structure.' He and Rosalind had a lot of arguments about the reflecting planes of the crystal structure, she throwing in her favourite expressions: 'Rubbish!', 'It's just not sound!', and 'It's not logical.' Gosling had trouble visualising the geometry of the arcs they were drawing, so Rosalind sent him out to buy big navel oranges which they used to simulate the spatial relations of the several curves. 'By the time we had finished,' said Gosling, 'there was quite a lot of orange peel on the floor.' There were also, in time, three serious papers in *Acta Crystallographica* to share their results with the scientific world.

On 18 July 1952 – with the air of freedom in her nostrils – Rosalind took her fountain pen and in capital letters wrote on a 3 x 6 inch card with a hand-inked black border a death notice for the DNA helix.

Wilkins was greatly upset by this prank. The practical jokes so enjoyed by many scientists were not his style, and about to go off to Brazil for a month to lecture on biophysics, he had no time for games.

To Gosling, Rosalind was simply playing devil's advocate. The 'death notice' referred only to the A form – 'crystalline' – of DNA. At no time, Gosling has said, did she believe the structure of the B form was anything other than helical. She could see how well Stokes's theories fitted her evidence. But she believed that it was impossible to prove the helix by model building and her arguments with Maurice were – so Gosling thought – an escape mechanism to ensure she was left alone to pursue the job to which she felt she had been assigned: to find the structure directly from the data, not from guesswork.

Stokes later admitted that he and Wilkins took the 'death' notice more seriously than Rosalind and Gosling intended. There was a 'memorial' meeting of sorts – a ceremony, a bit of an academic joke – but the spots were real. Rosalind had the Pat-

IT IS WITH GREAT REGRET THAT WE HAVE
TO ANNOUNCE THE DEATH, ON FRIDAY 18TH JULY 1952
OF D.N.A. HELIX (CRYSTALLINE)
DEATH FOLLOWED A PROTRACTED ILLNESS WHICH
AN INTENSIVE COURSE OF BESSELISED INJECTIONS
HAD FAILED TO RELIEVE.
A MEMORIAL SERVICE WILL BE HELD NEXT
MONDAY OR TUESDAY.
IT IS HOPED THAT DR. M.H.F. WILKINS WILL
SPEAK IN MEMORY OF THE LATE HELIX
R. E. Franklin RGGosling

tersons spread out all over the table. She and Gosling both wanted to display their evidence that offered an alternative explanation for the A form that was not helical.

Herbert Wilson, a doctoral student on a fellowship from the University of Wales, began working with Wilkins in the autumn of 1952, comparing DNA and nucleoprotein from the same source. He became aware of the unusual situation: Wilkins, who was responsible for initiating the DNA X-ray studies at King's and to whom Signer had given his precious DNA samples, was now excluded from studying it. Hearing Maurice mutter about Rosalind hoarding the best DNA for herself, Wilson asked the obvious question. Why not ask for some of the Signer DNA back? Wilkins strictly forbade him to do so.

There was no point. Rosalind was leaving on 1 January. Then Wilkins would have the Signer DNA all to himself. Wilkins continued to make regular visits to Cambridge. (Colleagues at King's had his fondness for Odile Crick as the motive.)

That term, disregarding Rosalind's warning, Pauline Cowan

also came to King's. There were not a lot of jobs available and she was interested in biophysics. She thought she had been engaged to construct a model of DNA as Wilkins was quite keen on the model-building approach. But she did not go ahead because Rosalind objected. The structure would not be solved that way, she said. As Cowan saw it, 'She was a very forceful personality, Maurice was not. So I got involved with collagen.' Cowan also formed the impression that Randall was essentially collecting people without a clear idea of what they were doing.

Nobody at the Cavendish in Cambridge was working on DNA. Bragg's ban still held. What Pauling was up to at Caltech was unclear. The Cavendish, however, now had a Pauling on the premises. Peter Pauling, Linus's younger son, was one of two fresh young American faces in Room 103. The other was Jerry Donohue, a physical chemist and former graduate student of Pauling's.

As a family, the Paulings were as close-knit as the Franklins, and more effusive. The parents, Linus and Ava Helen, corresponded regularly and lovingly with their second son, who had arrived in Cambridge, with a supply of colour film and a set of the *International Tables for Crystallography*, to live at Peterhouse, the oldest Cambridge college, to do research with John Kendrew. Their letters were usually about practical matters, such as the kind of car to buy when the elder Paulings came to Europe or whether Peter should wear woollen underwear, but there was a strong undercurrent of scientific information throughout. Peter, about to turn twenty-two, was struggling with the difficulty of entering a field in which his father was a world giant.

'Have you met Crick yet?' Linus wanted to know at the end of October. Indeed, Peter had, and liked Crick and Odile very much. They had a new baby and a picturesque small house called the Green Door in the heart of Cambridge and often invited Peter to Sunday lunch. Crick had edged into quasi-acceptance at Cambridge. The biologist Michael Swann had got him into Caius, as a member of the college, but not of the university; hence he could not wear a gown and was presumed to be a guest every time he exercised his dining rights.

Peter Pauling also became friendly with Watson, a young American like himself in pursuit of female company. He told his parents that he thought he would take French lessons, having learned from Jim Watson about a woman who had a house 'full of French girls'. 'No women in this town,' Peter explained. (So much for Newnham and Girton.) In his research, he was planning to use the new Cochran-Crick theory to determine the helical nature of protein molecules. He was also interested in the molecular structure of the myoglobin of sperm whales. 'Stranded whales are the property of the Queen,' Peter wrote home, 'but we have an agreement with her to get a piece of meat if one comes ashore.' In return, his father informed him that he and Corey had sent a paper to *Nature* proposing molecular structures for hair, horn and such substances, involving 'a parallel aggregation of helices'.

By November everybody at King's knew that Rosalind was leaving. Gosling was shocked and saddened. He had not realised that things had got that bad. Also he would be deprived of his thesis supervisor, right in the middle of work that was proving productive and publishable. There was so much more to do. However, her date of departure for Birkbeck had been postponed from January because she had lost a month's work with flu. She hoped to finish up the Patterson work for the two papers to *Acta Cryst* she was preparing on her DNA discoveries.

Her as-yet-unpublished results were given to the members of the MRC's biophysics committee when, on 15 December, responding to Randall's invitation, they came for their day at King's. Some of them, particularly their chairman, Sir Edward Salisbury, were uneasy about what 'Randall's Circus' was up to.

Like students preparing for the school's Open Day, each of the staff wrote up his and her current work; Rosalind no exception. She included the same information that was in her Turner and Newall report at the start of the year and handed it in to Randall who had it printed in advance, on 5 December. She gave the full dimensions of the unit cell, its length, width, and angles. She claimed to have established 'with certainty' that the crystal fell into the space group called 'face-centred monoclinic'. The results of a year's incomparable research, put like a gift in a box, for the MRC committee's holiday reading.

The committee came to King's at noon. First it was conducted by Randall around the new laboratories, then entertained to lunch. From two until four members were free to discuss with individual workers those parts of the research which interested them specially. Then they had tea (which had just come off the ration). The visit was judged a great success. The secretary of the MRC recorded that 'Salisbury [Sir Edward Salisbury, the chairman] is now satisfied that Randall and his team are concentrating their energies upon important problems.' An example mentioned was the use of the interference microscope to measure the dry weights of living cells and their components. 'This represented an important advance in quantitative cytochemistry, and it filled a need long felt by biologists.'

Clearly, King's work on the structure of DNA had not made a big impression.

The report circulated to the committee was not marked 'Confidential', nor was it confidential. On the other hand, in the customary British manner in which everything official is considered secret until deliberately made public, the report was not expected to reach outside eyes.

While King's was priding itself on having passed its MRC test, Peter Pauling in Cambridge received big news from his father. Pauling wrote that he and Corey had worked out a structure for DNA. Peter immediately showed the letter to Watson and Crick. Watson was devastated. He and Crick had long feared that Pauling might crack the problem before they did.

Watson left Cambridge to go skiing in Zermatt with his sister over the holiday break, convinced that Pauling had triumphed. He stopped off in London to break the bad news to Maurice Wilkins. Wilkins countered with the good news that Rosalind would leave King's as soon as she finished writing up her work and would not be taking DNA with her. He could then get down to model-building in earnest.

End of term at King's brought the annual Biophysics dinner, but Rosalind understandably chose not to attend. On Christmas Eve, Randall left his office and went along Fleet Street and up Ludgate Hill to St Paul's Cathedral. He climbed the many steps only to find the building locked. Outraged, he shot off a scathing letter to the editor of *The Times*, giving his address as his club, the Athenaeum. The letter appeared on 30 December under the heading 'Christmas at St Paul's':

> Sir – There can be few village churches throughout the land whose doors do not remain open on Christmas Eve for prayer, carols, and meditation, and in many churches Christmas Day begins with a midnight service. Why does the great Cathedral Church of St Paul, set as it is in the heart of London, fail in its duty in this regard? As I approached the cathedral about 5.15 p.m. on Christmas Eve, a peal of bells died away: but instead of the doors being wide open to receive those who wished to enter they were locked and many went bewildered away under the mocking lights of a Christmas tree which denied the spirit of Christmas . . . Cannot the Dean put his cathedral to better use at Christmas? No wonder the Church attracts such a small fraction of the public to within its doors at the one time of year when the Christian faith implies that they should have ever-open doors.

A more superstitious man might have interpreted the sacred portals shut in his face as a portent.

TWELVE

Eureka and Goodbye
(6 January–16 March 1953)

'Rosy, of course, did not directly give us her data. For that matter, no one at King's realised they were in our hands.'

James Watson, *The Double Helix*

THE SECRET OF LIFE, four billion years old, was unpicked in a drama that moved day by day, almost hour by hour, in the first seven weeks of 1953.

News that Linus Pauling and Robert Corey had solved the structure of DNA started the clock running. On 6 January Rosalind, who somehow had got wind of the news, wrote Corey at Caltech and asked for details. They had been in correspondence since May when he had marvelled at her 'splendid X-ray photographs of nucleic acid fibres'.

Her notebooks show intense activity as January progressed. She herself had started to think of building a model of the A form based on her Patterson calculations. Might these represent a figure-eight shape – two coils crossed in the middle? Or paired rods? She knew of Chargaff's ratios and tried to squeeze the four bases of DNA into a structure with the phosphates on the outside. They would not fit. Still bemused by the discrepancies between the A form and the B form, she accepted that the B form was a two-chain helix but still had doubts about the other. As her later collaborator Aaron Klug was to comment, 'The stage reached by Franklin at the time is a stage recognisable to many scientific workers, when there are apparently contradictory, or discordant, observations jostling for one's attention and one does not know which are the clues to select for solving the puzzle.'

She had been at work for a week when Peter Pauling returned to Cambridge from his winter break in Germany and Austria. On 13 January he wrote his father to ask for a copy of the Pauling-Corey paper on DNA, adding that the Cavendish's MRC Unit would like one too. He prefaced the request with a joke:

> You know how children are threatened 'You had better be good or the bad ogre will come get you.' Well, for more than a year, Francis and others have been saying to the nucleic acid people at King's, 'You had better work hard or Pauling will get interested in nucleic acids.'

Pauling sent the paper, confident that he had scored another victory over Sir Lawrence Bragg, his old rival at the Cavendish. In fact, he sent two copies of 'A Proposed Structure for the Nucleic Acids' to England, one to Peter and the other to Bragg. Peter replied, 'We were all excited about the nucleic acid structure. Many thanks for the paper. Second sunny day since I have been in England.'

The paper appears to have arrived on 28 January, and there was indeed excitement at the Cavendish. When Peter brought the paper to the lab, Jim Watson had to restrain himself from grabbing it out of Peter's hand. He held back, impatiently listening to Peter's summary, until he yanked it out of Peter's outside pocket and read it for himself. Instantly he saw that Pauling's proposed structure – a triple-stranded helix with the phosphates at the centre – was much like the mistaken model that he and Francis Crick had built in November 1951. Worse – or better, from Watson's point of view – Pauling had made a fatal chemical error. The phosphates were not ionised – that is, Pauling had not built in the electrical charges phosphates acquire when in water. What he was proposing as a structure for nucleic acid was not an acid at all.

Pauling had made the silly mistake because he was in a hurry. But why? There has been much retrospective speculation on why one of the world's greatest chemists, the holder of the Presidential Medal for Science, the author of the classic textbook on the nature of the chemical bond, should have risked his reputation by rushing into print with a carelessly flawed proposal. One suggestion was that Pauling, having cracked the structure of protein, wanted credit for solving the other half of the cell's secrets. Another is that, for all his many honours, Pauling had never won the Nobel prize. Pauling later said that his wife once asked him why he hadn't cracked the problem and that, upon reflection, his answer was: 'I guess that I always thought that the DNA structure was mine to solve, and therefore I didn't pursue it aggressively enough.'

Watson's delight in the error was tempered by the news that the Pauling paper would soon be published in February in the *Proceedings of the National Academy of Sciences*. The mistake would immediately be spotted and Pauling would be on the trail again. Watson felt he and Crick had about six weeks' breathing space.

Also on 28 January Rosalind gave her leaving seminar at King's. Maurice Wilkins thought she was long-winded; he strained to hear the word 'helix' and did not. Neither did Herbert Wilson, who took notes. She did not refer to the B form of DNA, nor show the superb Photo 51, but concentrated instead on the recent experimental work of herself and Gosling that suggested that the A form of the molecule was not helical.

For J.T. Randall, an unwelcome sight at the end of that month was Jim Watson. The gawky young American from the Cavendish seemed always to be turning up at King's College London. One morning Randall went in to the coffee club that met daily in Angela Brown's room and there was Watson, grinning. 'Here's the Dean of St Paul's!' he wisecracked. Randall's pompous Christmas letter to *The Times* had been the subject of general mirth in the lab.

No one else talked to the inventor of the cavity magnetron like that. Randall was furious. Once Watson was gone, he boomed out, 'Never let that man in my sight again!' Out of sight, however, was not out of mind.

Rosalind did not like the sight of Watson either. On 30 January the door to her office opened and in he came. The only published account of what ensued is his: a pivotal scene in *The Double Helix:*

> Since the door was already ajar, I pushed it open to see her bending over a lighted box upon which lay an X-ray photograph she was measuring. Momentarily startled by my entry, she quickly regained her composure and, looking straight at my face, let her eyes tell me that uninvited guests should have the courtesy to knock.

Watson asked her whether she wanted a look at Pauling's manuscript, and getting little response, rushed on to point out where Pauling had gone astray. She countered with her own evidence that a helical structure was by no means proven. But Watson had heard from Wilkins that Rosalind was 'definitely anti-helical' – neither of them having seen her evidence. (The Birkbeck crystallographer, Harry Carlisle, wrote in his memoirs, 'I am convinced from Rosalind's excellent X-ray studies on both the A and the B forms of DNA that she was not in the least "anti-helical" at that time as suggested by Watson in *The Double Helix*.') He felt that Rosalind was more concerned with extracting positive arguments from her X-ray data.' For his part, Watson decided that she did not know what she was talking about:

> I was more aware of her data than she realised. Several months earlier Maurice had told me the nature of her so-called anti-helical results. Since Francis had assured me that they were a red herring, I decided to risk a full explosion. Without further hesitation I implied that she was incompetent in interpreting X-ray pictures. If only she would learn some theory, she would understand how her supposed anti-helical features arose from the minor distortions needed to pack regular helices into a crystallising lattice.

> Suddenly Rosy came from behind the lab bench that separated us and began moving towards me. Fearing that in her hot anger she might strike me, I grabbed up the Pauling manuscript and hastily retreated to the open door.

'Fearing that in her hot anger she might strike me': the patent absurdity of this remark has caused much scorn. Rosalind was of slim build and medium height, Watson a stringy six feet plus. But the male fear of the female has always been absurd – the stronger afraid of the weaker – but no less real for that. To dismiss it is to dismiss the Medusa, the Loathly Lady, the Wicked Witch of the West and all the other guises for whatever the male resents and recoils from in the female; that led even the mild-mannered graduate student John Cadogan to say of Rosalind, 'She nearly terrified the living daylights out of me.'

Watson's own bewilderment with women is well chronicled in *The Double Helix*. He could not approach them unselfconsciously: they were either prey – 'popsies' or 'au pairs' – or goddesses such as his aristocratic Scottish hostess Naomi Mitchison or his sister Elizabeth. Rosalind was neither; worse, she was an angry woman. And she had reason to be angry. Courtesy might have demanded that Corey send a copy of their DNA paper to her, not to the Cavendish.

Watson portrays Rosalind's 'hot anger' as entirely unmotivated. There is another possible explanation for her rage – indeed of the whole incident. Early in 1953, very upset, she complained to a friend at King's that she had come back to her room one day and found her notebooks being read. If Randall and Wilkins saw themselves as her bosses, she stormed to her confidant, they should have protected her work better. Instead, she knew Wilkins to be in open and frequent communication with the Cavendish pair. She voiced her fears also to an old colleague from BCURA.

* * *

She herself was engaged in a race against time. Due to move to Birkbeck in the middle of March, she was rushing to finish up as much as possible of the Patterson interpretation of the A form of DNA before she left King's. She was working hard on three papers (to be published jointly with Gosling) so as to hand them to Randall before she left; his permission was needed before anything was sent out to other readers, let alone for publication. Two papers were for *Acta Crystallographica*. In them she was formally announcing to the scientific world what she had discovered at King's: the existence of DNA in two forms, and the conditions for readily and rapidly changing from one to the other. She described the DNA molecule. Its phosphates were on the outside, thus exposed and ready to take up water: making hydration, and therefore stretching, easy. Thus shielded in a sheath of water, the DNA was 'relatively free from the influence of neighbouring molecules'. She appended to the first paper the startlingly clear X-ray photographs she and Gosling had taken of both forms. The second paper gave the measurements on the x-ray pattern of the A form on which she and Gosling had concentrated in the past six months: full of the kind of information that crystallographers would appreciate. The third, a shorter paper, was a more general summary of their findings on the B form.

As Watson retreated from Rosalind's wrath, he was rescued from his fantasised assault by Maurice Wilkins, who had put his head round the door. Wilkins consoled him that some months earlier, 'she had made a similar lunge at him' and had blocked the door when he wanted to escape. As Wilkins poured out his 'see what I'm up against' tale of woe, he reiterated what Rosalind had said in her 1951 colloquium about the two forms of DNA and complained that he had been left to use only samples given him by Erwin Chargaff in New York, which would not produce the A to B transition. Rosalind's Signer fibres had produced much better patterns. 'She's got a very good B,' said Wilkins.

Wilkins had not known of Rosalind's excellent diffraction

photograph numbered 51, taken eight months earlier, until Gosling brought it to him sometime that January. Gosling, preparing to complete his thesis without Rosalind's supervision, had every reason to show what was also his own current work to the assistant head of the department. 'Maurice had a perfect right to that information,' Gosling said, looking back. 'There was so much going on at King's before Rosalind came.' Both he and Wilkins knew the DNA research would continue after she left.

Unguardedly, Wilkins showed Watson Photo 51. There were many diffraction photographs of DNA around the lab; this one was simply the best. As the information in it was not new to Wilkins – Rosalind had related many of the details in her symposium in 1951 – he had no idea it would strike Watson with the force of revelation. Nor did he have any idea that Watson was about to make a new stab at building a model of DNA.

But Watson was now wiser than in late 1951 when he had botched his first model. A year of working on the tobacco mosaic virus had educated him, as had reading the Cochran–Crick–Vand paper on helices. He was now able to sweep up at a glance the meaning of Rosalind's photographic image: unbelievably clear evidence of a helix, with detectable parameters of tilting and spacing. It was with little exaggeration that he wrote in *The Double Helix*: 'The instant I saw the picture my mouth fell open and my pulse began to race.'

The two men had dinner that night in Soho. What Wilkins wanted to talk about was Chargaff's ratios and the possibility that they might hold the key to DNA's structure. Watson, however, pressed for numbers to go with the pattern he had just seen. He extracted a few. The repeat (the length of one turn of the helix) of the B form was 34.4 Ångströms – ten times the spacing between the bases stacked 3.4 Ångströms apart. On the train back to Cambridge, Watson drew the pattern from memory on the only paper available, the margin of his newspaper. It was stark enough to fit in the small space. In his mind there still lurked the possibility that the molecule might have three helical chains. As he cycled

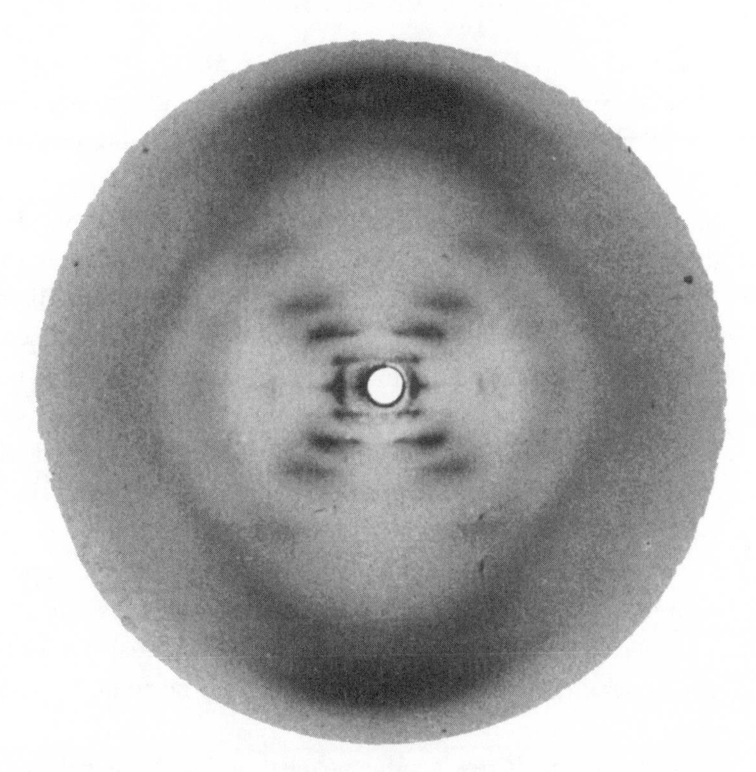

Rosalind's Photograph 51 (shown original size) of the B form of DNA, which told Watson that the molecule was a helix.

home from the train, mulling over Rosalind's photograph with its dark cross, he decided on two. 'Francis would have to agree,' he later wrote. 'Even though he was a physicist, he knew that important biological objects come in pairs.' (According to Crick, the argument was more complicated than that.)

Events then moved fast. The following day, 31 January, Bragg unleashed Watson and Crick. He consented to Watson's request to order metal components from the workshop to begin model-building again. Having Linus's DNA paper in hand, Bragg was not about to repeat the worst mistake of his life. The thinking at the Cavendish seems to have been, 'We missed out on the fibrous protein. Now Pauling is going to get the DNA as well.'

When Rosalind started her modelling attempt of the A form,

she found her paired-rod structures would not work. She then moved to a figure of eight in which a single chain formed a long column of repeating eights. But by Monday, 2 February, she ruled out the figures of eight as well.

On Wednesday, 4 February Watson started building. His sketch of Rosalind's Photo 51 confirmed to Crick that they had the right figures with which to put together a B-form model: 20 Å diameter, 3.4 Å vertical distance between the bases, repeat distance of 34 Å and helical slope of about 40 degrees. Although Crick was not entirely convinced that there should not be three chains, Watson was set on two. Once more the disputed phosphates formed the central core, ignoring Rosalind's argument in 1951 that these water-attracting groups had to be on the outside of the molecule. Crick has supplied the dialogue which shows teamwork in action: '"Why not," I said to Jim one evening, "build models with the phosphates on the outside?" "Because," he said, "that would be too easy" (meaning there were too many models he could build in this way). "Then why not try it?" I said.'

Watson obliged. Crick's other crucial action on 4 February was to invite Wilkins to come up to Cambridge for lunch on Sunday. Wilkins accepted, saying gullibly, 'I will tell you all I can remember and scribble down from Rosy.'

On Sunday, 8 February Wilkins arrived to lunch and found that the Cricks had two other guests, eager listeners: Watson and Peter Pauling. When the three men urged him to start model-building to get ahead of Linus, Wilkins vowed to do so as soon as Rosalind was clear of the premises. Watson and Crick wrung from him a grudging consent that they themselves could try again – without telling him that they had already started. Wilkins suddenly realised he was dealing with an awkward situation and left early for London. Had he said too much?

Perhaps he had, for sometime during the following week Watson and Crick got hold of the MRC report on the biophysics committee's December visit to King's. It was given them by their

Cavendish colleague Max Perutz, a member of the committee. Whether Perutz *volunteered* the MRC report, or merely handed it over in response to a direct request from Watson or Crick, has been the subject of some dispute over the years as its importance to their efforts has come to be realised. Defensively, Perutz wrote in *Science* in 1969, 'I was inexperienced and casual in administrative matters and, since the report was not confidential, I saw no reason for withholding it.'

The MRC report was all Watson and Crick could hope for – as valuable as an enemy's code book. The section on Rosalind's work explained the change from the first to the second type of DNA structure, and specified, with a neat little table, the measurements of the 'face-centred monoclinic unit cell' 'with certainty'.

One glance, to Crick's well-stocked brain, translated Rosalind's information into a recognition that the crystalline form of DNA belonged to the space group called 'monoclinic C2'. The DNA crystal, in other words, looked the same when turned upside down. Crick instantly knew, therefore, that one chain of the two helices must run up and the other down; they were anti-parallel, like up-and-down escalators. Watson took some persuading, but had to accept Crick's better judgement: 'monoclinic C2' was exactly the space group Crick had been working with for his thesis on horse haemoglobin.

During the same week, on 10 February, Rosalind wrote in her notebook: 'Structure B: evidence for a 2-chain (or 1-chain) helix?' She did some calculations on the photograph numbered 49, then laid the B form aside for two weeks. She had her two long *Acta Cryst* papers to finish and get to the typist. (These must have been completed shortly after, for they were received by the journal's English editor on 6 March, for publication in late summer.)

The paper by Pauling and Corey was published by the National Academy of Sciences on schedule in February. Their proposal for

DNA's structure may have been all wrong, but their explanation of DNA's importance was excellent:

> The nucleic acids, as constituents of living organisms, are comparable in importance to proteins. There is evidence that they are involved in the processes of cell division and growth, that they participate in the transmission of hereditary characters, and that they are important constituents of viruses. An understanding of the molecular structure of the nucleic acids should be of value in the effort to understand the fundamental phenomena of life.

To cover their British flank, Pauling and Corey also sent a letter to *Nature,* which appeared on 21 February, announcing their structure. From that, if Rosalind had not already gleaned the fact from Watson's traumatic visit three weeks earlier, she could see that Pauling had misplaced the phosphates. She wrote to Pauling and told him so. It took courage for a thirty-two-year-old 'Attached Worker' to correct a world authority but she was sure of her facts. She explained why her data indicated that the phosphate groups had to be on the outside rather than at the core of the molecule. She also mentioned, as if talking to a friendly associate, that she had three papers on DNA underway awaiting Randall's approval.

Pauling remained convinced that he was right. He was working without the fine-focus X-ray equipment or the pure DNA that Rosalind had, relying on Astbury's pre-war X-ray diffraction patterns, on which the A and B forms were superimposed. Nonetheless, he politely promised to try to meet Rosalind when he came to England (having succeeded at last in getting his passport back). However, he confessed in a letter to Peter that when in London he would much rather see Pauline Cowan, whom he had met at Oxford. He did not remember ever having met Miss Franklin but he had heard the previous summer that she was leaving King's; Corey had a very good opinion of Wilkins. Linus passed on to his son the information that Rosalind was working on three papers on DNA. As for his own structure, he could see that it was a 'tight squeeze' but he and Corey were checking it over and expected things to come out all right.

Two chains, yes, but what held them together? If the phosphates were on the outside of the molecule, the four bases always found in DNA would have to be squeezed inside, somewhere between the chains. The problem was that each of the four – the two purines, adenine and guanine, and the two pyrimidines, cytosine and thymine (often abbreviated to A and T, G and C) – had a different shape.

Mid-February saw Watson wrestling with the problem. He tried various combinations, using cardboard cut outs because the metal pieces had not yet arrived from the machine shop. First, on 19 February, he tried pairing like with like: the resulting pairs were either too small or too large for the diameter specified in Rosalind's MRC table. Then he tried joining a purine to a pyrimidine – A to T, C to G. Still they didn't fit.

At another desk in Room 103 was Pauling's former student, Jerry Donohue. Like watching someone fumble with a jigsaw, Donohue suggested that Watson was using the wrong pieces. Bases might exist in two forms – that called 'keto' makes an oxygen atom available for bonding in a different position from that in the contrasting form, 'enol'. Why not use the 'keto' form? Donohue asked. Because, Watson answered, in effect, J.M. Davidson's *The Biochemistry of Nucleic Acids* and other textbooks showed the enol forms. The textbooks were wrong, said Donohue. As a chemist, he passed the news on to Watson that 'enol was *out* and keto was *in*'.

On the last Monday of the month, 23 February, Rosalind probably after finishing her *Acta* papers, took out Photograph 51 and began her careful measurements of the B form. By the following day, she had accepted that the A and the B forms were both two-chain helices. 'Nearly home,' wrote Aaron Klug in his analysis of her notebooks years later.

It occurred to her that a way of explaining the Chargaff ratios – why DNA invariably contained the same number of A as T molecules and the same number of Cs as Gs – was that adenine and guanine were interchangeable, and so were cytosine and thymine. This interchangeability gave her a vision that 'an infinite variety of nucleotide sequences would be possible to explain the biological specificity of DNA'.

'Base interchangeability is, of course, a long way from the final truth of base pairing,' Klug acknowledged. She was two steps from the solution. Crick later ventured that she would have taken those steps within three months.

She did not. She failed to understand the significance of the monoclinic C2 space group or to figure out the bases joined in pairs to carry the genetic code. 'It is easy to feel sympathy with Franklin,' the science historian Horace Freeland Judson has written. 'The fact remains that she never made the inductive leap.'

But Rosalind had been trained, as a child, as a Paulina, as an undergraduate, as a scientist, never to overstate the case, never to go beyond hard evidence. An outrageous leap of the imagination would have been as out of character as running up an overdraft or wearing a red strapless dress.

The fact is that in two unhappy years, working in isolation except for Gosling, in a field new to her, she had come within two steps of answering the most exciting question in post-war science. What is more, she, unknowingly, had provided all the essential data for those who took the two brilliant leaps of intuition – to anti-parallel chains and base pairs – that cracked the problem.

Donohue's off-the-cuff advice about the 'keto' form tipped the balance for the Cavendish. Watson, fitting together his pieces of cardboard at the end of the month, suddenly noticed that adenine joined by hydrogen bonds to thymine formed the same shape as cytosine bonded to guanine. The paired pairs were congruent; Chargaff's ratios were explained. What is more, the same pairs were always found together. Adenine always grabbed thymine;

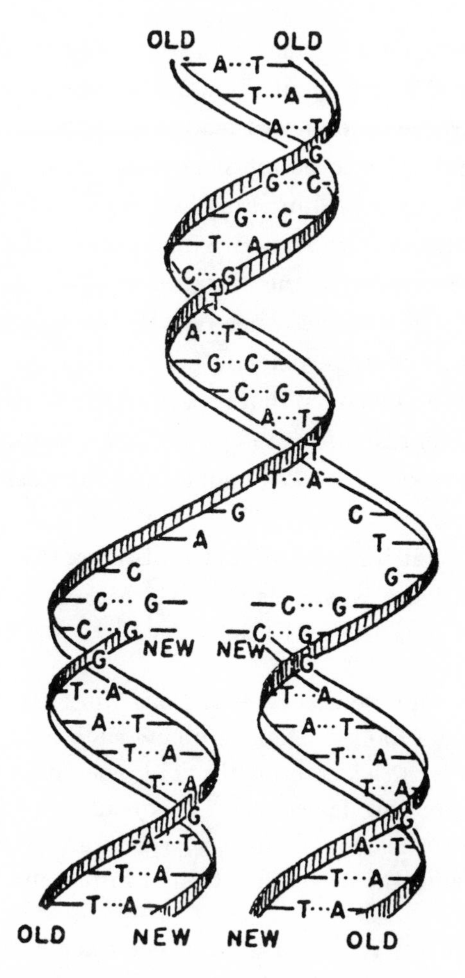

How the double helix copies itself. The two sugar-phosphate backbones are like the rails of a spiral staircase, with the paired bases, adenine and thymine (A and T) and guanine and cytosine (G and C), forming the steps. When the rails come apart, each of the bases seeks its invariable complement, and thus two staircases are formed, the new identical to the old.)

cytosine, guanine. Whenever any one of them appeared along the long DNA chain, the other was invariably across from it on the opposing chain. One side was therefore the inverted mirror image of the other. Split apart, each of the chains served as templates on which a new, complementary chain could be formed. Base pairs, therefore, held the secret of DNA – how characteristics pass from generation to generation, from cell to cell, from parent to child. The DNA molecule is a set of chemical instructions which reverses the tendency of matter to become disordered and allows new molecules to be the same as the old. The day was 28 February 1953. It was at lunchtime when (according to Watson's myth-making narrative) 'Francis winged into The Eagle and told everybody we had found the secret of life.'

Donohue later was one of many who felt that his part in the great discovery was underplayed. 'Let's face it,' he wrote in 1976, 'if the fates hadn't ordained that I share an office with Watson and Crick in the Cavendish in 1952–53, they'd *still* be puttering around trying to pair "like-with-like" enol forms of the bases.'

The patronising attitude of the Cavendish team towards Rosalind continued. As Peter wrote his father in March about the rivalry between the Cavendish and King's:

> Wilkins is supposed to be doing this work; Miss Franklin is evidently a fool. Relations are now slightly strained due to the Watson-Crick entering the field. They have some ideas & shall write you immediately. It is really up to them and not to me to tell you about it . . . They are getting pretty involved with their own efforts, and losing objectivity.

Never had Maurice Wilkins's gift for metaphor served him better. Exhilarated by the imminent departure of his nemesis, on 7 March, a Saturday, he wrote to Crick:

> I think you will be interested to know that our dark lady leaves us next week . . . I am now reasonably clear of other commitments and have started up a general offensive on Nature's secret strongholds . . . At last the decks are clear and we can put all hands to the pumps!
>
> It won't be long now.
> Regards to all
> Yours ever
> M

It was too late. Nature's secret strongholds had fallen to the pitchmen from the Cavendish. The final model of the double helix of DNA (or at least one turn of it; one entire DNA molecule, reproduced on that scale, would have been 50,000 miles high) had been finished on Saturday, 7 March. Crick was looking at it as he read Wilkins's letter.

Rosalind could write a strong letter too. On 10 March, with no idea of what had happened at the Cavendish, she poured her heart out to Adrienne Weill in Paris, saying that her prime concern was to get out of King's as fast as possible.

> I start at Birkbeck next week. It got delayed repeatedly – first because I missed a month with flu and things in the autumn, then because I thought another month would give me a lot more results – but it didn't, and I'm abandoning an unfinished job now in order to get out of King's without further delay. I don't know yet what I shall do at Birkbeck. In theory I do what I like, which probably means I shall waste a lot of time playing around with odds and ends before I settle to anything. As far as the lab is concerned, I shall be moving from a palace to the slums, but I'm sure I shall find Birkbeck pleasanter all the same.

On 12 March Wilkins made the journey to Cambridge to see for himself what he had learned on the telephone from Kendrew, Watson and Crick apparently being too embarrassed to tell him directly that they had cracked DNA. There stood the model, its elegance proclaiming its accuracy. Wilkins (unfortunately for Rosalind, King's being over-modest yet again) spurned any idea of joint authorship: 'I felt the model *as such* was their work. Crick agreed with alacrity.' Separate publication was decided upon. Wilkins then returned to London and, in his words, 'I told *everybody*.'

For Rosalind, the news that the Cavendish had cracked DNA was an irrelevant parting gift. She had confidence in her own two papers, on their way to publication, that would give solid data on DNA's structure, not a mere hypothesis. In any event, she and Gosling had their shorter paper on the B form nearly ready and she would turn her hand to polishing it up the following week.

Rosalind came to say goodbye to Freda Ticehurst and to thank the photographer for all she had done for her. She said (in Freda's recollection), 'I'm not wanted here – we [meaning Wilkins] could never work together. It's impossible for me to stay.' Her valediction delivered, she packed up, and turned her back on King's College London. One thought that certainly never crossed her mind was that half a century later the institution she despised would name a building after her, coupling it with the name of the adversary who had branded her forever as the 'dark lady'.

THIRTEEN

Escaping Notice

ROSALIND'S PART in the great discovery was obscured by a series of manoeuvres made behind her back. For Watson and Crick, once they had solved the structure, the next step was to publish quickly. Speed was essential for staking a claim. A short letter to *Nature* was the obvious medium. But they were in a tight spot. They knew they were right to declare their inspired insights. But how to give credit to the King's experiments that had been the springboard to their imagination? They could hardly refer to Rosalind's data which had not yet been published. An explicit acknowledgement would publicly reveal the Cavendish's intrusion into King's territory, the violation of the gentleman's agreement, as Watson put it, 'of not working on the problem of a close friend'. A third snag was the Medical Research Council report. Officially unpublished, the report could not easily be cited as a reference without Perutz being identified as the transmitter. In any event, before the Cavendish duo could do anything, the heads of the laboratories concerned, Bragg of the Cavendish and Randall of King's, would have to come to an accommodation on acknowledgements.

When (according to his recollection, it was 12 March 1953) Maurice Wilkins went up to Cambridge to see the model, Watson detected 'not a hint of bitterness in Maurice's voice', rather a recognition that the structure would be of great benefit to biology'. The following day Wilkins tried to telephone Crick and could not reach him. History can be grateful, for Wilkins penned another of his vivid letters, combining disarming modesty with stoic acceptance of fate:

> Dear Francis,
> I think you're a couple of old rogues but you may well have something. I like the idea. Thanks for the MSS. I was a bit peeved because I was convinced the 1:1 ratio was significant & had a 4 planar group sketched and was going to look into it & as I was back again on helical schemes I might, given a little time, have got it. But there is *no good grousing* – I think it's a very exciting notion & who the hell got it isn't what matters.

Whatever his true feelings, Wilkins knew his moral advantage. Who the hell got it *did* matter. So did the glaring fact that the experimental work had been done entirely at King's. He pressed Francis gracefully, apologising for 'being a trifle awkward on 2 points'. First, he thought Bruce Fraser should have an opportunity to publish the ideas he had proposed in late 1951 and had built into a reasonable (if three-chain) model:

> we rather stopped him doing so a year ago because we thought we could do better and didn't. I have sent him a cable. I don't think Fraser's thing anything like as nice as yours but it's a hell of a sight better than Pauling's.

Wilkins, wanting credit for the considerable work he himself had done on DNA, then added that he would like to publish a brief accompanying note in *Nature* at the same time, with a picture showing the general helical case. He made no mention of Rosalind's work.

As he was finishing his letter to Crick, Wilkins was interrupted by Gosling saying that he and Rosalind had a paper all written. As indeed they did; Rosalind had told Pauling about it a couple of weeks earlier. 'Just heard of a new entrant in the helical rat race,' Wilkins added to Crick, uncharitably interpreting the news as Rosalind 'jumping on the bandwagon': 'So at least 3 short articles in *Nature*. Christ!' Wilkins signed off with another metaphoric flourish: 'As one rat to another, good racing. M.'

The mood at King's was in reality not quite so light-hearted. Randall was, in the words of Willy Seeds, 'like a scalded rat' when he heard that the Cavendish had beaten King's. Geoffrey and Angela Brown felt that the news was a very big blow to Wilkins and that he was devastated. Gosling felt 'quite upset, quite scooped'. Anthony and Margaret North remembered that 'everyone had forty fits at King's when Cambridge got there first. Randall was furious, his nose out of joint. After all, he had created the biggest biophysics laboratory in Britain and had little to show for it.'

'Had it been the other way around,' the irascible Jerry Donohue said later, 'if someone anywhere had done the same with the data collected by the MRC group at the Cavendish, the resulting eruption would have paled that of Krakatoa to a grain of corn popping.'

Randall was in no position to erupt. Instead, he made sure that his lab would have two of its own papers in *Nature* along with Watson's and Crick's and that theirs would be held up until those from King's were ready. 'Get writing!' his order went out. L.J.F. Brimble, moreover, one of the two editors of *Nature*, was, like Randall, a member of the Athenaeum, and, as the first of many to feel sorry for King's having been pipped at the post, gave Randall the opportunity to have King's work represented in what was going to be a momentous occasion.

In the ensuing panic to get King's papers ready, Wilkins cabled – a dramatic gesture in those austere times – Bruce Fraser in Australia and asked him to write up his model as quickly as possible in a note suitable for publication in *Nature*. Fraser complied. In view of the urgency, he sat up all night typing the very paper Wilkins had suppressed two years earlier and sketching the diagrams. There were no photocopying machines in 1953, and had Fraser not done it by hand, the manuscript would have lain several days in a queue before being photographed for duplication. Then, even more expensively, Fraser had his work cabled off to London next morning.

But when the Fraser paper arrived, Crick vetoed it: what was the point of publishing wrong data? Instead, he and Watson appended an acknowledgement to Fraser at the end of their own paper, describing the Fraser model as 'rather ill-defined' and the Fraser paper as 'in the press'. It was not, and was never published.

In their own historic letter to *Nature*, Watson and Crick referred to 'the following communications', meaning the two papers from King's – 'Molecular Structure of Deoxypentose Nucleic Acids' by Wilkins, Alec Stokes and Herbert Wilson; the other, 'Molecular Configuration in Sodium Thymonucleate' by Franklin and Gosling – with the comment: 'We were not aware of the details of the results presented there when we devised our structure . . .'

That was the truth, narrowly speaking. Even though they had not seen the King's papers when they sent their own first draft to Wilkins, Watson and Crick did know many of the details of the King's work. They knew these from Maurice's conversation, from Rosalind's photograph and from the MRC report – enough to admit, in guarded language, that all the experimental work had been done at King's, and in a paper written a year later to say, in a footnote, that 'without this data [from the King's College Group] the formulation of our structure would have been most unlikely, if not impossible.' In the excitement of discovery it seems to have escaped their notice that while Rosalind's work was fundamental to the discovery, she had not been consulted on its use. But once they had seen the beauty and simplicity of the model, the importance of her data receded. They hadn't needed it; once they had grasped the base pairs, the copying mechanism was clear.

Had L.J.F ('Jack') Brimble and Arthur Gale, joint editors of *Nature* since 1939, known of the true genesis of the Watson-Crick discovery, they might have insisted that Rosalind's name be included among the principal authors of the primary paper. Or they might have positioned her and Gosling's paper, with its vital photograph, second, rather than third, in the sequence of three papers. Sadly, the editors' part in the orchestration of the whole tripartite publication in *Nature* will never be known. The voluminous and unsorted *Nature* archive was thrown out during a move to new offices in 1963; thus there was nothing available when Macmillan, the publishing company which owns *Nature*, sold its own archive to the British Library in 1966.

Did Rosalind suspect the part her work had played in the design of the double helix model? In retrospect Gosling, Perutz and Crick believe that she must have done. It was obvious once she looked at it that the dimensions of the molecule fitted her data on the diameter and the repeat number of nucleotides in a turn. Yet she did not complain. Insofar as is known, she did not protest to anyone, not even to Gosling, who went up with her to Cambridge to see the model. Her attitude, as he recalled, was 'It's very pretty, but how are they going to prove it?'

In the event, there was a certain amount of pre-publication negotiation between London and Cambridge about the actual wording of the three papers. During the editing, Wilkins himself played a part in toning down Watson's and Crick's delicately crafted admission of the information they had obtained from King's. Wilkins asked Crick to cut out the word 'beautiful' from his text (possibly a compliment paid to Rosalind's and Gosling's beautiful X-ray photographs). 'Delete *very beautiful,*' he suggested, 'and say "We have been stimulated by the work of the group at King's or something."' He also asked: 'Could you delete the sentence "It is known that there is much unpublished experimental material." (This reads a bit ironical.)'

Nothing is more ironical than the words Rosalind wrote, inserted by hand into her typescript in order to adapt it to accompany the Watson-Crick paper. The alteration transformed her own fundamental findings into a 'me-too' effort. The inserted words were: 'Thus our general ideas are consistent with the model proposed by Crick and Watson.' So indeed they should have been consistent, considering that the Watson-Crick model was in large part derived from her work. The striking Photo 51 of the B form of DNA appeared as an illustration to her and Gosling's own paper, with no suggestion that Watson had seen it, let alone been inspired by it. She appended also her comment that the photograph 'is strongly characteristic . . . of a helical structure'.

Watson and Crick's short classic letter glides to its end with the penultimate, now-celebrated, line: 'It has not escaped our notice that the specific pairing we have postulated immediately suggests a possible copying mechanism for the genetic material.'

It had not escaped Watson's and Crick's notice either that Rosalind's work was fundamental to their discovery and that they had not consulted her about its use. While Rosalind seems to have suspected that her data had somehow reached Cambridge, no one at King's nor at the Cavendish bothered – or, considering her temper, dared – to press the matter further. To all concerned, she was an awkward member of King's staff on the way out.

Randall himself may never have realised exactly what had happened although he must have been half-aware of Wilkins's role as transmitter. In any event, he was unconcerned with justice for Rosalind. He certainly gave Rosalind no thanks for her contribution to the great discovery. There was a party at King's when the three papers came out in *Nature* on 25 April 1953, under the umbrella heading, 'Molecular Structure of Nucleic Acids.' During the celebrations, the photographer Freda Ticehurst looked around and asked pointedly, 'Where's Rosalind?' In reply, she got 'some looks'.

Far from thanking her, Randall sent a letter to Rosalind at Birkbeck a week before the trio of papers on the structure of DNA were to appear in *Nature*. Addressing her formally as 'Miss Franklin', he told her not only to stop working on DNA, but to stop thinking about it:

> You will no doubt remember that when we discussed the question of your leaving my laboratory you agreed that it

would be better for you to cease to work on the nucleic acid problem and take up something else. I appreciate that it is difficult to stop thinking immediately about a subject on which you have been so deeply engaged, but I should be grateful if you could now clear up, or write up, the work to the appropriate stage. A very real point about which I am a little troubled is that it is obviously not right that Gosling should be supervised by someone not specifically resident in this laboratory. You will realise that the necessary reorganisation for this purpose which arises from your departure cannot really proceed while you remain, in an intellectual sense, a member of the laboratory.

Yours sincerely
J.T. Randall

It can be argued that scientific discovery is not creativity in the sense that artistic composition is. 'Science differs from other realms of human endeavour,' according to Dr Walter Gratzer, 'in that its substance does not derive from the activity of those who practise it.'

If Beethoven had not written his Ninth Symphony, no one else would have done it. In contrast, if Watson and Crick had not discovered the double helix of DNA, others would have found it, and probably not long after. Rosalind came close to getting the answer herself. That her draft manuscript dated 17 March 1953, proving how close she was, should only have come to light many years later is one of the many ways in which fortune did her no favours.

Part Three

FOURTEEN

The Acid Next Door

(March–December 1953)

IN MOVING TO Birkbeck College in Bloomsbury, Rosalind was, as she said, moving 'from a palace to a slum', but it was a slum teeming with interest. The world-renowned crystallographer J.D. Bernal, the head of the physics department, appreciated that she was a brilliant experimentalist. He was eager, as he told his administrative superiors, 'to get her assistance in a new attack on problems of virus structure, for which we have particular experience and facilities here'.

There were no 'hooded crows' at left-wing non-sectarian Birkbeck. The college was founded in 1823 as the 'London Mechanics' Institution' in much the same spirit as Ellis Franklin's 'Working Men's College': to offer evening education to craftsmen and small employers 'in the principles of the arts they practised and in the various branches of science and useful knowledge'. It became Birkbeck in 1907 and a full college of the University of London in 1920. Located in Bream's Buildings near Chancery Lane, the college held all its classes at night; a condition of admittance was that the student be engaged in full-time day employment.

The college barely survived the rupture of the war. In 1951 a dreary new modern block on Malet Street replaced a building lost during the war, adding to the University of London's architectural desecration of Georgian Bloomsbury. For the rest, Birkbeck operated in scrappy makeshift accommodation, with its science done under appalling conditions. Crystallography teaching and research were done nearby at 21, 22 and 32 Torrington Square, in bomb-damaged red-brick eighteenth-century houses more suitable for film sets than laboratories. Within the *Upstairs, Downstairs* interior, scientists worked on structures of viruses and searched for life forms in meteorites. Outside, a descendant of Robert Louis Stevenson's 'Old Lamplighter' came on a bicycle at dusk with a burning rag on the end of a long pole to ignite the gas lamps, returning in the morning to put them out.

Rosalind's office was on the fifth floor in what used to be a maid's room, reached by a narrow staircase leading up from the main floors below. The X-ray equipment was in the basement, the former kitchen, where the ceiling leaked. She had to hold an umbrella over her head while setting up her equipment.

John Desmond Bernal was so much a legend in his lifetime that he lived comfortably with the nickname he had acquired as a Cambridge undergraduate, 'Sage'. He knew, and everybody knew he knew, everything. He was equally at ease in French and *Finnegans Wake*. His omniscience inspired many stories. It is said that when the *New Statesman,* preparing a profile, asked Bernal to name a subject with which he was unfamiliar, he suggested 'eighth-century Roumanian churches'. A few days later he asked if he might amend that to 'sixth-century Roumanian churches'. In C.P. Snow's Cambridge novel *The Search*, Constantine, a scientist modelled on Bernal, is shown walking home from a meeting of the Chemical Society, discoursing on the Cretan Renaissance and Chinese economics in the Tang dynasty.

Bernal's background was as eclectic as his interests. Born in Ireland to a Sephardic Jewish father of Catholic faith and an American Presbyterian mother who converted to Catholicism and who spoke French to him as a boy, Bernal was educated by the Jesuits. Moving on to Cambridge he became a scientist and a Marxist. His field was crystallography and he taught, among others, Max Perutz, Maurice Wilkins and Dorothy Hodgkin. During the Second World War he was scientific adviser to Lord Louis

Mountbatten, Chief of Combined Operations. Mountbatten much admired him, Communist affiliation notwithstanding, and took him to the Quebec Conference in August 1943 to help plan the D-Day landings. Bernal was helped in choosing the landing beaches by the Anglo-Norman poem, 'Roman de Rou' about the escape of Duke William (not yet the Conqueror). This provided old Norman place names which, when studied, revealed a silted-up harbour. For his patriotic service to the Allies, in 1947 Bernal was awarded the US Medal of Freedom. Further American honours and indeed an American visa were denied him after 1949 when he became vice-president of the World Peace Committee, making frequent trips to Eastern Europe. In 1951 he founded Scientists for Peace, the forerunner of the Campaign for Nuclear Disarmament. In 1953, just as Rosalind joined his lab, Bernal eulogised Stalin in an obituary, called him 'a great scientist' as well as 'the greatest figure of contemporary history'.

Bernal had left Cambridge before the war and accepted the chair of physics at Birkbeck in 1937; he was elected a Fellow of the Royal Society the same year. At Birkbeck he introduced the use of crystallography to study plant viruses and very large molecules with molecular weights of millions, (hydrogen having a molecular weight of two; oxygen, thirty-two.) Trying to revive this work after the war, he set up the Biomolecular Research Laboratory, opened in 1948, to resume the study of virus structure abandoned at the beginning of the war. He wrung a grant of £61,100 from the Nuffield Foundation for biomolecular research. Birkbeck contributed a smaller sum. It took over the funding in 1952 and tucked the work into its physics department, where Werner Ehrenberg and Walter Spear developed the high-intensity X-ray tube that enabled Rosalind at King's to photograph her single DNA fibres.

Bernal wanted a separate department of crystallography but was blocked by the Master of Birkbeck, John F. Lockwood. Their uneasy relationship was to plague Bernal, and by extension, Rosalind, during her Birkbeck years. Lockwood felt that the physics department (thanks to Bernal's fame) was 'becoming unbalanced'.

At fifty-two, Bernal's passions were crystallography, women and the Soviet Union. He attracted a stream of celebrities with left-wing political views to his flat at the top of 22 Torrington Square. The most celebrated, both of whom came in 1950, were Pablo Picasso (who left a large wall drawing behind) and the singer Paul Robeson, who sang – an event commemorated in a plaque that, like the Picasso drawing, adorns the college's drab concrete.

Success as a Don Juan was part of the Bernal legend. Far from handsome, he had an unruly shock of hair and terrible teeth, but a roguish smile, dazzling intellect and a fascinating stream of talk accompanied by an intent gaze on his quarry. His endless conquests were bolstered by an ideological commitment to sexual freedom, tolerated by his wife, whom he married when he was twenty-one and still at Cambridge, and by the two mistresses by whom he also had children. His reputation gave rise to many Bernal jokes: (Q: Why are there so many women in crystallography? A: Because Bernal was a crystallographer.) Another arose from Bernal's long collaboration in the 1930s with the American scientist Isidore Fankuchen. The faithful Mrs Fankuchen, so the story goes, was so amused by Bernal's scorecard that she started a society, 'Women Who Have Not Been to Bed with Bernal.' She was president and Bernal's secretary, Anita Rimel, was treasurer. One day after the Fankuchens had returned to Brooklyn, Mrs Fankuchen is said to have received a telegram saying 'You are now President AND Treasurer of the Society.'

The great Dorothy Hodgkin was more than a conquest: her long affair with Bernal began about 1934 after she had left Cambridge. Between 1933 and 1936 they produced twelve joint crystallographic papers.

Bernal encouraged contacts with the Soviet Union and influenced *Acta Crystallographica* to print its instructions to authors in Russian as well as English, in the hope of eliciting papers from the other side of the Iron Curtain (as he did not call it). He remained well-connected, at the same time, with the British Secret Service. One of his recruits to the staff, Dr John Mason, was asked to keep an eye on Bernal by New Scotland Yard, which he did. Yet if Mason ever reported anything to MI5, Bernal knew about it the next day.

Rosalind stayed aloof from Bernal's politics. Nor was she drawn into his amorous activities, because, said one of her staff, 'she never gave him the slightest encouragement'. Part of his brilliance was attracting brilliance – the reason why Francis Crick had tried to work with him after the war. Bernal was also good at mobilising research grants and at delegating. He set up a number of very good research groups at Birkbeck and left them to themselves, while he travelled the world. For Rosalind Bernal was an understanding and supportive boss. Having directed her to a tough new subject, he let her read into it and otherwise left her alone. He told the college that while Miss Franklin was already constructing special apparatus for crystallographic research on viruses, 'It is now clear, however, that she cannot expect to start effective observations until the autumn.' She was therefore free to continue to interpret the X-ray diagrams of DNA and their Patterson functions.

Much of her first months was spent in meetings with Gosling. Laughing in defiance of Randall's ban, they worked on their joint papers and discussed his thesis. Perforce she remained 'in an intellectual sense' (Randall's phrase) part of King's laboratory, for Gosling had not been given another thesis adviser. For this service Gosling rewarded her with an effusive solo acknowledgement in his introduction to the thesis. Ignoring the work he had done with Wilkins and Stokes, Gosling wrote: 'I am deeply indebted to Dr R.E. Franklin, who has introduced me to the techniques of X-ray crystallography and with whom I have worked most closely throughout this investigation.'

Rosalind laughed off Randall's letter telling her to stop thinking about the nucleic acids as 'Just the sort of thing they do there.' She could hardly stop thinking about the nucleic acids. Ribonucleic acid (RNA) was an important ingredient of the tobacco mosaic virus to which Bernal was directing her.

RNA was the acid next door, a second form of the nucleic acid found in every cell. It is like DNA except for having one more oxygen atom attached to each of its sugars. Jim Watson said of his own useful research in 1952, when Sir Lawrence Bragg had ordered him off DNA, 'I had decided to mark time by working on tobacco mosaic virus (TMV). A vital component of TMV was nucleic acid, and so it was the perfect front to mask my continued interest in DNA.'

Rosalind needed no front. All the work she and Gosling had done with the fiddly Beevers–Lipson strips had not been wasted. In addition to the two major manuscripts sent to *Acta Cryst*, which had been accepted for September but needed adjustments, they composed a second note for *Nature* that spring, giving their evidence for 'a two-strand helical molecule of the Structure A form of DNA' and saying that it was 'based mainly on a study of the cylindrically symmetrical Patterson function of Structure A'. The note appeared on 25 July, and confirmed the Watson–Crick proposal in principal but not on 'points of detail'.

Rosalind was always eager for invitations to speak, especially abroad. She liked always to have two speaking engagements in her diary, to reassure herself – she told a friend – against the day 'when it is all over'. That day seemed hardly imminent. In April she went to Aachen to read a paper on 'The Mechanism of Crystallite Growth in Carbons' (later published in German) and in June to Paris, to read a short paper on 'Le rôle de l'eau dans l'acide graphitique'. This paper was full of plural references to work done with her old mentor, Jacques Mering: 'Nous avons étudié les diagrammes de rayons X', 'nous déduisons' and 'A partir de nos résultats' but, as before, only her name appeared on it when it was published. While in Paris, she was invited to do some work: 'So I've been back to the old lab, with the same people and same apparatus – all very pleasant.' Among the French treats she brought back to London from her satisfying journey was a curved piece of quartz crystal to make a focused X-ray beam.

(Her hospitable bent moved her to loan her flat whenever possible. During her frequent travels, she kept 22 Donovan Court full of friends passing through London and, on one occasion, when a London friend found herself temporarily without accommodation after having a baby, Rosalind moved out and installed mother, grandmother and infant in the flat, which she had stocked with a cot and nappies.)

In the spring of 1953 Rosalind could not have dreamed what Watson, Crick and Wilkins would be saying from platforms in the twenty-first century – that her contribution was critical to the discovery of the double helix of DNA. With no idea that her data had been used, she had no sense of having been overtaken in a race which the Cavendish had won. Rather, Watson and Crick were now her collaborators, on the way to becoming her friends. In any event, she did not accept their DNA structure as more than a hypothesis. She referred to it as 'the Watson-Crick model' and in the second of her *Acta Cryst* papers, both of which appeared in September, she wrote that 'discrepancies prevent us from accepting it in detail'.

While her *Acta* papers were in the press in the spring and summer of 1953, she sent copies to Watson for his comments and received a reply from Crick, saying that as Jim had left for the States, he would answer instead. Addressing her as 'Miss Franklin' (the correct form for writing to a female colleague; a man would have been addressed as 'Franklin'), Crick asked – as if their own historic letter to *Nature* had not settled the matter – whether the phosphates in DNA really must be accessible (i.e., on the outside); whether calf thymus was the only source to yield Structure A and what fibres gave only Structure B? About her second paper, he wondered whether the unit cell was '*truly* face-centred monoclinic, and not really triclinic, with two angles 90\167/'. The point was important, he said, 'because if the unit cell is *strictly* C2, one must have the DNA chains in pairs, running in opposite directions'.

Rosalind's paper does indeed say 'C2 is the only space group possible'. This shows that, even though she had missed its anti-parallel implications, she understood very well what the space group C2 was. It is a puzzle why Watson and Crick, considering the information so crucial to their discovery, had not mentioned the fact in their two *Nature* papers of 1953. Nor did they mention the space group by name in the much longer, more detailed paper on the DNA structure that appeared in the *Proceedings of the Royal Society* the following year. This omission, viewed half a century later, remains inexplicable.

Crick was not being ingenuous in his letter to Rosalind. The questions he posed were still open. The letter to *Nature* of 25 April 1953, a turning point in scientific history, had not struck scientists of the time like a thunderclap from heaven. Their awakening to its importance was very gradual. Many did not believe it for years, Erwin Chargaff remaining a conspicuous and sarcastic hold-out. Other scientists recall their initial scepticism.

At the Institut Pasteur in Paris, François Jacob recalled, the Watson-Crick article 'had not electrified me or anyone else in the laboratory. The crystallographic argument went over my head.'

When Gunther Stent, at the Virus Laboratory at the University of California at Berkeley, heard the news from Max Delbrück in March 1953, he ran to tell his colleagues that there had been a fabulous breakthrough. They weren't interested. 'Look at Pauling!' was their reaction. 'It'll be just a few weeks until some biochemist is going to show that the Watson-Crick structure is baloney as well.' But Pauling was not quick to admit his ideas were 'baloney' and continued, even after he read the Watson-Crick paper, to believe that his own three-chain structure had merit.

Crick himself much later recalled that he had certainly believed their structure was:

> along the right lines (though Jim had occasional doubts), but the experimental evidence, though supporting the model, was at that time not enough to prove it beyond a reasonable doubt. Strictly speaking, our model was not finally *decisively* proved till some 25 or so years later, when the crystal structure of short lengths of DNA of defined sequences was solved by the isomorphous replacement method.

What remained to be discovered was how the double helix came apart. The two chains could not copy themselves without unwinding, but no one knew how it was done. In July 1953 the Royal Society's summer *conversazione* included among its exhibits a model and presentation entitled 'A proposed structure for DNA', which identified seven scientists as contributors: Watson, Crick, Wilkins, Franklin, Wilson, Stokes and Gosling. Gosling was guarding the model when along came the eminent J.B.S. Haldane, 'puffing on this foul Woodbine, and he looked at it for a long time, and then he said: "So what you want is an un-twiddle-ase".' (In biochemistry, '–ase' is the suffix meaning an enzyme.)

The importance of the discovery, however, was recognised at Cold Spring Harbor where Watson turned up at a symposium in June 1953. The triumph was celebrated in the inevitable show, in a ditty for two voices, one announcing 'I'm Watson', the other, 'I'm Krick':

> Let us show you our trick –
> We have found where the seed of life sprang from.
> We believe we're a stew
> Of molecular goo
> With a period of 34 Ångström.

Another verse hit at the well-known girl-watching proclivities of the intrepid pair:

So just think what this means
To our respective genes –
That sex need not be disgusting;
All you girls try the trick
Of Watson and Krick
And achieve double helical lusting.

A Radcliffe student, waiting on tables at the symposium, fed up with the flirtatious banter, said she had named her two cats Watson and Crick. 'And,' she added darkly, 'they're neutered.'

So slow was the double helix news to travel that when in the summer of 1953 Bernal introduced Rosalind to Sven Furberg, whose pioneering work at Birkbeck in 1949 was cited in the Watson-Crick paper, Furberg was surprised to find a copy of his thesis on Rosalind's desk. She asked him whether he believed the Watson-Crick proposal that the DNA molecule was a helix. *Believe* it? He had not even read it. (Bernal later reproached himself for the road not taken and thought if Birkbeck had pursued Furberg's work, there might have been an 'almost simultaneous discovery'.)

Waiting for the apparatus in order to get down to serious work on the tobacco virus, Rosalind had time for a prolonged holiday. Like her Aunt Mamie Bentwich, Rosalind had always been sceptical about the Zionist dream of a national homeland for the Jews. Now that the dream had been realised and Israel was a five-year-old independent state, she decided to go and take a look.

Boarding the 'Tauern Express' at Victoria Station, with Anne Piper as company as far as Greece, she headed first for Ljubljana where she was received once more with great warmth. A scientific colleague said to Anne about Rosalind, 'She makes my clock tick.' Anne returned to England from Greece and Rosalind got a boat to Haifa from Piraeus.

At Haifa she was met by relatives and friends, who included her cousin Irene Neuner, her wartime housemate in Putney. They organised for Rosalind to be taken by taxi to Jerusalem where she toured ten synagogues and witnessed tight orthodox communities. Her ambivalence about Jews was stirred. She wrote her parents, who, she knew, would understand:

> I find it hard to see what can be the common factor in anti-semitism which is directed both against the ghetto Jews and the successful assimilated or near-assimilated Jews – so different that I find it hard to remember that what I saw last night was the way the majority of Jews lived in Poland and other East European countries right up to the last war.
>
> I find it repulsive that the young should be made equally grotesque, and deliberately isolated from anything resembling a broader life and forbidden any of the normal subjects in school . . . I'm told that these young – who finally get away – frequently are fantastically anti-religious or communist.

She had no such doubts about the Weizmann Institute of Science at Rehovoth. Her second cousin David Samuel, grandson of Herbert Samuel, was working there: another scientist in the family. She found the research excellent and the climate wonderful. She might, she told her mother, 'be tempted to seek work there: but the community was too small and isolated – all 28 and 35 and having babies at the same time . . .'

As she went about the country, however, the sight of kibbutz life disturbed her deeply. Identifying, as so often, with the children, she wrote home:

> I can sympathise with the desire of somebody whose intellectual background is fully developed to go off to the country and let their intellect thrive on more rudimentary things (though I should never want to do it myself) but for children who have never been elsewhere or even seen town life – it can't mean the same thing.

Determined to see as much of the country as she could, she hitchhiked to remote spots – something her cousin Irene felt was a very risky thing for a young woman to do. At some point a lorry driver who gave her a lift tried to rape her. Her cousin later commented 'I think it shook even her'; but Rosalind was undeterred.

She went to the southernmost tip of Israel, to Eilat on the Gulf of Aqaba; the town then barely existed – it was just a handful of huts and a wind pump. She also made her way to Sodom on the Dead Sea. More difficulties appeared.

> The only 'restaurant' was a large cave by the Dead Sea [run by a] man who sold bottled drinks . . . I was escorted to the cave by some French-speaking Moroccans and ate sardines and drank orange juice. There seemed to be no option but to spend the night in the cave, but a party of tourists arrived and took me back to Bersheba – 6 men from 4 different countries.

Only towards the end of her stay did Rosalind visit Tel Aviv. Perhaps it was just as well. Her disapproval of Jews who fulfilled the stereotype flooded over her, as she explained to her parents:

> I had been more than adequately warned about the people, who disgusted me far more than the architecture. I'm afraid German Jews in Tel Aviv make just the same impression on one as Germans in Germany, typified by the hotel keeper: 'Mein Gott, warum you don't speak Deutsch.' They all cheat you as badly as orientals – and one minds it more . . . In fact, Tel Aviv would make anyone anti-semitic.

Back in Haifa, relating her experiences to Irene, Rosalind told about her near-rape. Her cousin privately thought, with the sarcasm of a relative, 'Ros wouldn't know what being raped was.'

Irene knew Rosalind well enough to ask her directly what no one else dared: why had she never married? Rosalind gave a straight answer: all the men she had ever liked were always already married. Irene silently reasoned that Rosalind was not the type for breaking up a marriage.

None of the awkward moments was relayed to her mother,

for whom Rosalind summed up the holiday, in her last air letter on 9 September: as 'wonderful . . . amazing, varied'. Back at Birkbeck, over cucumber sandwiches, she gave a talk to the Fabian Society at which she described her trip to Israel, particularly the kibbutzim and the research she had seen at the Weizmann Institute. Bernal was so impressed that he took steps to have a visa for Israel put in his passport, while Wolfie Traub, a member of the Birkbeck crystallography group, was so intrigued that he joined the Weizmann Institute a few years later and spent most of his scientific career in Israel.

Her intended work on the tobacco mosaic virus was still being held up. Bernal, trying to lessen the delay, asked Randall if they might borrow from King's 'one of the cameras designed by Miss Franklin for the study of X-ray diffraction by solution'. Randall's reply was true to form: 'We do in fact intend to use this equipment very shortly and I shall have to make careful enquiries before I can give a definite answer one way or the other . . .'

At last, by December, Rosalind had redirected, without apparent regret, her scientific curiosity and ingenuity to the tobacco mosaic virus, known from the curling, brittleness and mottled patches of light and dark green (hence 'mosaic') that it causes on tobacco leaves. To Rosalind TMV was just as exciting as DNA.

Viruses are large particles made up of inert large molecules composed of proteins and either RNA or DNA. They come to life when they enter a living cell and parasitically take over the reproductive mechanism of the cell and duplicate themselves rapidly, with consequences well-known as 'getting a virus'.

Research on TMV was not motivated by a desire to assist the then-respected tobacco industry, but rather by a wish to understand the way it infects – injecting itself like a syringe into the host cell. TMV virtually launched the science of virology in 1886; it was simple, stable, available and highly infectious – a model for the study of viruses in general. Before the war Bernal at Cambridge, with his colleague Fankuchen, had put TMV under

What TMV (tobacco mosaic virus) does to tobacco leaves.

X-ray diffraction analysis. Together – known as 'Sage' and 'Fan' – they produced a classic paper in 1941, showing that the virus was composed of identical subunits of protein; how these fit together was anybody's guess. Watson got the answer in 1952 by photographing a tilted sample held before the X-ray camera. This exercise was very far from his expertise, yet he succeeded in establishing that the matching subunits spiralled round in a helix. This accomplished, Watson gave up the work, concluding that 'the way to DNA was not through TMV'. For Rosalind, on the other hand, TMV was the way out of DNA. Rosalind built on Watson's work, correcting it with the help of her X-ray photographs, better ones than, as with DNA, anyone had got before.

In her attic office, with her North American Phillips camera five floors below in the leaking basement, she set out to determine the internal structure of the helical array, using the techniques in which she was singularly adept. She wrote to Dr A. Lindo Patterson in Pennsylvania to tell him, what others in the field already knew, that TMV was going to be harder to solve than DNA: 'Here the fibre-diagram is considerably more complicated than that of DNA, but judging by previous work (I've only just taken it up myself) it is not possible to index the reflections . . .'

The questions to answer were: did the nucleic acid, the RNA, stand in the centre, like the wick in a candle? Or was it tucked into cosy corners between the subunits?

The momentous year of 1953 ended with a healing visit to Nannie at Church Stretton and to the Luzzatis in Strasbourg. Denise and Vittorio, after a year at the Brooklyn Polytechnic Institute, had returned to France, both for posts at the Centre des Recherches sur les Macromolécules in Strasbourg. Rosalind summed up the year in a Christmas letter to Anne and David Sayre in Philadelphia:

> For myself, Birkbeck is an improvement on King's, as it couldn't fail to be. But the disadvantages of Bernal's group are obvious – a lot of narrow-mindedness, and obstruction directed especially at those who are not Party members. It's been very slow starting up there, but I still think it might work out all right in the end. I'm starting X-ray work on viruses (the old TMV to begin with) and I'm also to have somebody paid by the Coal Board to work under me on coal problems – more or less the continuation of what I was

> doing in Paris. But so far I've failed to find a suitable person for the job . . .
>
> Really the only interesting thing that's happened as far as I'm concerned was a wonderful summer holiday in Israel this year.

It was just as well that Rosalind was out of earshot of King's annual Christmas entertainment, in which a big theme was not that King's had been outwitted by the Cavendish but that Rosalind had gone. In the programme's 'advertisements' there was one for 'Rosy's Parlour':

> Best crystalline NUCLEIC ACIDS:
> Dehydrated, Uviated, de-Rosieated
> (Frustrated Export), Infraredded, X-rayed
> And otherwise Maltreated
>
> Also: Hands Read, Bumps Told, Patterson Diagrams
> of your Future (and Past)
> by Request
> Madame Raymonde Frankline
> Clairvoyante

But King's was far from Rosalind's mind. She had had an exciting year, quite apart from going to Israel and helping to discover the double helix of DNA. The next one looked like being even better. She had just received her first invitation to visit the United States.

FIFTEEN

O My America
(1954)

CARBON BROUGHT ROSALIND to the United States. With no hint of interest in her work on DNA, in 1954 the Gordon Research Conferences, an annual summer series on scientific research held in the green hills of New Hampshire by the American Association for the Advancement of Science, invited her to attend its August session on 'Coals and Related Substances'.

If she could raise the money, that is. The Gordon organisers could not cover travel expenses for foreign visitors 'except for the cost of getting from the port of entry to New Hampshire and back'. Nor did the Bank of England lightly permit the export of sterling. For the would-be visitor from Britain the challenge was to find sponsors to pick up the bill, and Rosalind wanted to see a lot more than New Hampshire. An essential stop would be the new Virus Laboratory, headed by the Nobel laureate Wendell Stanley, at the University of California at Berkeley.

The Gordon organisers did try to help Rosalind find the money. They told Rosalind they could guarantee her about $700 if, in addition to attending the coal conference, she were willing to give lectures at Pennsylvania State University and at laboratories in Pittsburgh and Cleveland. For its part, Britain's National Coal Board wrung permission from the Bank of England to provide Rosalind with £250 on the condition that any unexpended balance be refunded on her return. As she began scraping together other contributions to pay her way to the West Coast,

Rosalind was much helped by Watson and Crick. In April Crick, working for a year at the Polytechnic Institute of Brooklyn, sent her encouraging news (she was now 'Dear Rosalind'):

> Jim and I talked to [Wendell] Stanley. He seemed interested (about T.M.V.) and suggested he could pay part of your expenses.
>
> I was very pleased to hear about your T.M.V. photos. Jim is certainly not doing any further work on T.M.V. but is concentrating on R.N.A. His old idea for T.M.V. was a helical arrangement of *globular* molecules (more or less) and as far as I know he has not developed this further.
>
> Yours ever,
> Francis.

Rosalind duly wrote begging letters to Wendell Stanley and also to Linus Pauling:

> Since leaving the work on D.N.A. in King's College I have been working here on tobacco mosaic virus . . . As you know, the dollar exchange position makes it impossible to obtain funds over here for such a journey, and I wondered if it would be possible for me to give a lecture or so in return for a fee which would cover a part of my expenses. Arrangements so far take me as far as Cleveland (Ohio) first week in September.

Pauling replied that, while two weeks before the new term began was not a good time for a lecture, they would make an exception for her. He offered an honorarium of $50, plus covering the costs of her stay at the Athenaeum – Caltech's faculty club.

Writing to a physicist at Yale who was interested in X-ray diffraction work on DNA, she allowed herself a more detailed explanation of what exactly she was doing. Sending reprints of her two *Nature* DNA papers, she said she had begun:

> some similar work on tobacco mosaic virus. I have got X-ray fibre-diagrams with more than 300 sharp maxima and hope to have a cylindrical Patterson function calculated before

> the summer. I think there is considerable support for J.D. Watson's suggestion that the rod-shaped particle is a single helical molecule; and if this is the case, then the cylindrical Patterson is a particularly powerful method of investigating it and there is a good chance of its showing up some of the main structural features.

The American adventure started badly when she was refused a visa. In February the consul at the US Embassy in London turned down her application in the belief that she was going for the purpose of earning money: 'There is considerable objection to granting a visa to anyone coming to the United States to be gainfully employed in competition with American labor.' Four months of correspondence with the US Bureau of Mines and with the National Carbon Research Laboratories in Cleveland were necessary before the visa was granted on the understanding that Rosalind would give back any funds in excess of her expenses. The Cleveland scientist in charge of her arrangements apologised for 'the unfortunate atmosphere which prevails in Washington' which he saw as having clouded her application.

Naturally, Bernal approved of her trip. The United States, the world centre of scientific research, was closed to him, a banned person, not least because his own travels that year were taking him to interview Khrushchev and Mao Tse-tung. Any suspicion that Rosalind herself was seriously left-wing is dispelled by her lack of an FBI file. The Bureau's file on Bernal is encyclopaedic.

Rosalind was having money troubles at home too. Her Turner and Newall Fellowship was to run out in December, and what then? In mid-1954 Birkbeck asked the Agricultural Research Council to give her a three-year contract for plant virus research and for assembling a team of assistants. The appropriate salary, said Birkbeck, would be £1,100 a year, but the ARC reduced the salary, with no explanation, to £1,080. The cut was small but hurtful. It was the beginning of a difficult relationship.

Rosalind was up against an administrative wall. Academic

staff had tenure and job titles. Researchers such as Rosalind, largely supported by outside financing, were 'external staff', with lower status. Rosalind had no formal teaching duties, although as director of a research group she did a great deal of teaching in an unobtrusive form.

How unobtrusive this could be – and an example of her extraordinary dexterity – was discovered by a Birkbeck student taking a crystallography practical examination in the early summer of 1954. 'One would see from time to time a serious dark-haired lady but one did not know whether she was a student, a technician, or a member of staff as she had no link with the course.' On the evening of Martyn Pease's practical exam, she was the invigilator, looking, he thought, bored and scornful of their efforts. Each student had one hour to take an oscillation X-ray photograph, which required positioning a needle-shaped crystal onto a strand of fibre glass, sticking it down with glycerine, and then setting it upright in the dead centre of the camera. Pease was next to last. 'The pitfalls were many,' he recalled 'and I fell in every one.'

> The crystal I chose was too big and its weight caused it to slide down the fibre slippery with the glycerine. Time passed rapidly and it was ferociously hot. I was verging on panic. Eventually fibre and crystal were on the camera stage and I attempted to get the needle vertical and on the axis of rotation of the stage. Much as I fumbled with the adjusting screws it wasn't coming right. Rosalind was leaning against the wall nearby. I turned and said quietly, 'Is this straight?' She looked through the eyepiece and wordlessly turned the knobs once each slowly in the correct direction. It was a matter of seconds. It was only as she stepped back that she realised she had committed a misdemeanour and stalked off in a cold fury. I pressed the start button and developed the film when the oscillations ceased. I won't say that it was fit for the textbooks but the examiners would clearly pass it on the nod.

* * *

Rosalind made warm-up trips in May to Zagreb, where she gave a lecture on 'Some aspects of the ultra-fine structure of coals and cokes' organised by the Croatian Chemical Society and Croatian Physical Society; she also went to the Julian Alps where she climbed to the top of Triglav (a 2,896-metre peak). In June she went to Paris for the Third International Congress of Crystallography, where she gave three papers, one with Mering. She enjoyed also introducing some of her new scientific friends from Yugoslavia to French and British colleagues, taking them to cafés on the Left Bank and organising a picnic at Fontainebleau. Then Rosalind was ready for America. With the exception of the visit to Israel, all her travels so far had been within Europe and she had never expressed interest in going to the United States. Like many British socialists, she distrusted America's prosperity, materialism and Cold War politics, and she disliked what she knew of loud-voiced Americans from the GIs during the war and tourists in London and Paris. For an intrepid traveller, however, the prospect of a 3,000-mile journey was exhilarating.

In 1954 the United States seemed very far from Europe. Transatlantic flights from London stopped at Prestwick, Reykjavik, Gander and Nova Scotia. As the first in her family to visit the United States, Rosalind sent home full reports, which were essentially travel diaries. She began on the plane:

> Stopped one hour for refuelling in Iceland – dull and populated by the American Air Force. To my great regret I missed seeing Greenland and icebergs, through trying to sleep. Lunch on plane. Stopped again at Gander! here the second lunch was called dinner. The reaction of all airlines to delays seems to be to produce food at frequent intervals on the assumption that this will more than compensate passengers for the time lost. Real sleeping facilities at stopping points would be far more appreciated.
>
> Gander was pleasant and sunny, and we were quite sorry to be pushed back into the plane. Newfoundland, from above, is most weird – apparently uninhabited, a mixture

> of scrubby grass rock and water (in all shapes and sizes from puddles to large lakes). As you approached the sea, the density of ponds and lakes increases, till gradually water predominates and the sea takes over.

Arriving in Boston eleven hours late, Rosalind lost a day's planned sightseeing. She had time to register only a brief first impression: 'the architecture is quite as ugly as London's but the streets are so much wider that the general effect is almost "continental". And it is certainly warmer.'

In New Hampton, New Hampshire, the New World began to make itself felt. The conference lectures were held in the morning and evening, leaving the afternoon free for outdoor pursuits. She had never thought, as she wrote Colin and Charlotte, 'of America as having trees, hills, clouds and sky more or less like Europe'. She laughed at her mother's worry that she was 'alone in America'. Far from that, she said, she was 'living with about 100 people in a school in a remote and green part of the country'.

> The overwhelming impression so far is of overabundance of everything and the resulting complete self-confidence of individuals . . . Superficially they have everything and too much of it. This part of the country consists of high hills and forest, the district being depopulated because it's not enough to support the American standard of living . . .
>
> Meals are enormous, and consist mainly of protein and ice. I'm beginning to understand why they serve sweet pickles, maple sauce, raw onion, cranberry jelly, stuffed olives etc. with all main dishes. It's because at some stage there is a process for extracting all the taste from the main dish . . . But the general effect, with added spices, is quite pleasant.

Her prejudices about American accents remained: 'I was puzzled by someone talking about the crocodiles she'd eaten, then discovered she'd said "hot dogs."'

The Gordon Conference was on her old subject, coal, and the assembly more industrial than intellectual. Not until she got back to Boston did she enter the scientific mainstream. Her first laboratory visit was to MIT, which, she explained to her parents, meant Massachusetts Institute of Technology and was 'actually a University with about 10,000 students'. There she met Alexander Rich, a specialist in nucleic acids who had done work on photographing TMV, and was in transition from Caltech to the National Institutes of Health in Washington. He offered to drive her to her next stop, the Marine Biological Laboratory at Woods Hole.

Before leaving, Alex Rich and his wife Jane arranged for Rosalind to stay overnight at the home of Jane's mother in Cambridge near Harvard Square. Mrs King, from an old Cambridge family which had sent seven generations to Harvard, did not meet her English guest the night she arrived. In the morning, however, she cooked and served Rosalind a full breakfast. On leaving the house, Rosalind quietly asked Jane if she should tip the maid.

At Woods Hole she was given a room, or rather a bed, at a family of summer residents who had rooms to let. The Littles, known locally as the 'little Littles', had eight children under twelve. Soon after Rosalind arrived, the older children announced that there was going to be a hurricane. They were right. Rosalind observed the developing drama with scientific detachment:

> It certainly was an interesting day. The wind had been strong all night. Look at the map to see that Cape Cod is a long and very narrow peninsula jutting straight out into the Atlantic south of Boston. Woods Hole is the near end of the peninsula, but with sea very nearly all round. When I first went out, about 9 a.m., the wind was lifting large amounts of sea and blowing it across the village, and the water level, 3 hours before high tide, was already well above the normal high tide level. The wind was very strong but I've known English gales reach similar strength. I think that the speciality of the hurricane wind is its persistence in the high strength and constant direction so that the sea piles up far above its normal level. Soon we saw from the laboratory

> windows that piers and landing stages were breaking up.
>
> Then the water came up over the sea wall – just rising steadily, with a rough sea but not really giant waves. Cars parked in the lower parts of the town began to get wet and there was a rush to find high parking places. I came back to see how things were in my digs and found water half way up the walls in the sitting room and the older children proudly told me that they'd been giving their younger brother a swimming lesson in the sitting room. We got back to the labs – the most substantial building here – just before they became isolated from the rest of the town by water 5 or 6 feet deep. There was an endless succession of strange sights. Boats broken loose were trying to get in at closed windows. Large landing stages, boats and dustbins drifted around the streets – one of the biggest rafts came to rest against a submerged car about ¼ mile from the sea. In fact it was all exactly like the picture that one sees of heavy floods but doesn't really expect to become involved in. The rise of water was fastest between about 12 and 1, then, quite suddenly, the water receded – without any comparably sudden change in wind speed. This left a new succession of weird sights – unexpected objects perched in unexpected places. A large part of the afternoon was spent trying to move a yacht from the main road. The mast had broken and the top half, attached by steel hawsers was tied in complicated knots around the telephone wires. While a large group worked with knives, poles, hooks, etc., a uniformed coastguard stood by repeating, 'This is a highway, the road must now be cleared at once!!' Much of the rest of the day [was] spent in sweeping water out of houses, making fires to dry houses, and spreading out wet objects. (My own things stayed dry on the 1st floor). There were no casualties in this area.

At Woods Hole she was also greeted by Jim Watson, arrived from Caltech. Celebrity had begun its descent on Watson. In the August issue of *Vogue,* his face appeared on the same page as Richard Burton's, described – Watson's, not Burton's – as having 'the bemused look of an English poet'. For Jim the glamour made up (nearly) for Crick having been invited to explain the double helix in a forthcoming edition of *Scientific American.*

To Watson's relief, Rosalind was very pleasant and eager to talk to him about TMV. He realised that her recent X-ray diffraction pictures confirmed and extended his 1952 proposal that the protein subunits of TMW were arranged in a helix. He also had to agree with his friend Leslie Orgel (or so he wrote half a century later) 'that she had been judged most unfairly when working on DNA'.

With apparently no fear that she might strike him in her hot anger, Watson offered her a lift to California as he and Sydney Brenner, a South African microbiological research student en route to Caltech from Oxford, were about to make the transcontinental drive. But she had to decline. She had tight travel plans with many stops, mainly in the industrial belt of the mid-West, for her lectures on coal, before she could think of the West Coast.

When the time came to move on, Rosalind was sorry to leave Boston:

> which I was beginning to think of as my home town on this continent. Boston prides itself on being a cultural centre and really to some extent is so. The public library is beautifully laid out around a large pseudo-Italian courtyard and fountains, with seats around where people sit and read and keep cool on hot evenings. There is even one-half of the courtyard reserved for non-smokers.

From Woods Hole she went on to New York, which she found uncivilised and rude, with important exceptions: the museums and the skyline, seen from the Brooklyn hilltop flat of the David Harkers of the Brooklyn Polytechnic with whom she was staying. (Harker was a protein crystallographer.) The three of them, Rosalind said, spent most of the time on the terrace roof with a view across to Lower Manhattan: 'The sky-scraper group is surprisingly beautiful, and changes with changing light like mountain scenery. I have to keep telling myself that it's real.'

In Philadelphia she saw Anne and David Sayre, resuming an old friendship and seeing another great art museum. She still nurtured her disapproval of American plenty, but it was beginning to melt. She wrote Colin and Charlotte:

> I'm also astonished by the amount of everything to spare – especially good land unused, even just around Boston, because there's better land elsewhere or more money in New York. The greater part of New England is just wasted. It makes nonsense of all the world-planning talk about cultivating the desert and the jungle – all they have to do is to cultivate America and distribute the products, and world food production problems would be solved.
>
> So far I've met 2 very different groups. The most industrial and research people at New Hampton, quite pleasant but not very interesting, working in large numbers with vast amounts of money and getting no further than Europeans with much more limited means. But in the other field, fundamental biology, America is really leading and they have a remarkable number of first class people among the under 30s. In spite of the vast distances, they all know each other rather well, and constantly exchange visits.

With some pity, she said that someone had told her that England had a hot day and she hoped it had come while Colin and Charlotte were at the seaside. 'Here, apart from hurricanes, the weather is wonderful.'

In a letter to her mother she was again full of praise for American scientists. She said she had met 'a surprising number of really first class people who form an elite group superior to any I have come across in England'.

Her conversion had begun.

Rosalind crossed America, sought after, singing for her supper. She gave her coal lecture at least six times. For the first part of her trip she was usually surrounded by 'the carbon crowd' of respectful industrialists:

> My lab visits were in 2 curiously interesting groups. In the carbon world the work I did in Paris seems to form the background to a large part of the industrial research going on here, and I'm welcomed as an 'authority' on the subject. So much so, that I had a nightmare in which I was surrounded by solemn middle-aged Americans all saying 'and what we should like to ask your opinion about next Miss Franklin is . . .' and then I woke up. In the biological laboratories, on the other hand, I have much to learn and almost nothing to give, and it is a question of hunting out the people who enjoy wasting their time talking to me.
>
> The contrast between the two groups of people I meet is also very striking. The carbon crowd are entirely uninspired, but in the biological laboratories there are an impressive number of really first-class people.

'First-class' again: the Oxbridge seal of approval. After less than two weeks in America, she wrote to Bernal that already she had heard of three serious X-ray diffraction studies on TMV and wondered if she could quickly write some papers when she got back, even before he returned from China.

As she worked her way across the continent, seeing many eminent scientists – Erwin Chargaff, George Gamow, Vladimir Vand, Isidore Fankuchen – some of whom she had met in Europe, she formed her impressions of their surroundings.

'Pittsburgh is as black as Manchester or Newcastle . . . though the inhabitants all insist that it got black *before* the new smoke-control laws were introduced. However, the blast furnaces just outside the town pour out a frightful orange-red smoke, presumably nitric oxide, such as I've never seen anywhere.' She was not impressed by Pennsylvania State in western Pennsylvania. It was 'a university of 11,000 students in a town of population 12,000 built in "campus" style, well-built and laid out, but devastatingly dull. I can't think how even the natives can tolerate living there.'

Rosalind's 'Letters from America' are a fine contribution to the tradition stretching from de Tocqueville and Fanny Trollope to Alistair Cooke. Dazzled by the sweep of the United States, with new impressions crowding in faster than she could write them down, she composed her letters home with the enthusiasm of a Victorian traveller and the precision of an experimentalist. (She expected them to be circulated, even to Nannie, then kept for her return – as they were.) In Chicago she described being shown round the city by the Ellises, the former Eisenstadters whom the Franklins had helped escape from Austria in 1939. They took her to the art museum and to the theatre: 'a comedy – heavy-handed in the American manner'. Travelling to Cleveland she got a good look at Lake Erie, which:

> seen from the laboratory windows, was bright blue. But the water is badly polluted by industry, and repulsive to look at close up. I was driven to and from the airport by a scientist who, in the usual American way, told me all the details of the mortgage on his suburban house, and the money he'd borrowed to buy a car.

American homes and gardens fascinated her. She was taken to a faculty home in Madison, Wisconsin, where:

> The front door opened in to an enormous living room, the only one in the house, as in medieval England (no door). In one corner is a huge fireplace, encased in stone in which was a fine blazing wood fire (the evening was chilly). The principal difference between the Americans and their ancestors is that the former use open fires for decoration only.

Halfway across America her prejudices had moderated. She had discovered Americans to be hospitable, practical, charitable and art-loving, even if their cultural appreciation left something to be desired:

> In their museums . . . they've got whole pillaged European churches and cloisters, and if anybody had told me about it before I should have been furious. But these things are beautifully laid out, and with the brilliant sunshine one gets here one can really for a moment get something out of the

> atmosphere of medieval Italy. Not only do I feel that this must give the poor Americans some compensation for living so far from Europe, but I would almost think it is worth sending some more cloisters as a civilising influence.

From Chicago she went to St Louis, to the botanical laboratory of Dr Barry Commoner at Washington University of St Louis. Commoner was working on an abnormal protein he called B8; it was not a virus component but resembled the protein in TMV. For Rosalind, this was the most interesting visit so far and they began what would be a fruitful collaboration which would result in a joint paper to *Nature* the following June.

As she proceeded westwards, she was increasingly dazzled (as who is not?) by the grandeur of the desert, the mountains, the clear air, the red vegetation and exotic birds. Rosalind neglected nothing. She journeyed to the bottom of the Grand Canyon by mule train and deplored the use of 'the Indians as tourist attractions . . . which I find rather horrible'. She did not mind the other residents dressed as if in a cowboy movie and joked about an isolated ranch where 'I looked in vain for a hero to gallop in on a white horse'. She liked American trains too; its third class was more comfortable than European first class. Never shy of a generalisation, she classified Americans for her parents. They were at best:

> kind and well-meaning but incredibly ignorant and self-satisfied. On the other hand, the university scientists are as fine a crowd as I've met anywhere. They throw themselves into their work with a whole-heartedness that is rarely found in England and get real pleasure out of it . . . And they seem very good at working together happily and unselfishly in groups in a way that is rare in England.
>
> Today I complete the transcontinental journey – the weather is heavenly, with a wonderfully clean atmosphere.

In California, she first visited the University of California at Los Angeles (UCLA) and was pleased to see that 'The oldest influence in Southern California – going back some 150 years – is Spanish, and this has a pleasing effect on architecture in the better parts, such as the stuccoed arches and red tile roofs.' She had some good scientific conversations with virus and botany researchers, including Sam Wildman, professor of botany, and Albert Siegel, a geneticist, then moved on to Linus Pauling's institution, the California Institute of Technology (Caltech), at Pasadena 'about 25 miles away (but still in Los Angeles)'. She gave two talks, the first of which Pauling attended, on X-raying crystals of tobacco mosaic virus.

At Caltech she met Jim Watson again. There was much to talk about, such as the news from a young crystallographer at Yale, Don Caspar, that the TMV virus had a hollow centre, filled with water. If so, where was the RNA, the nucleic acid that had been assumed to be there? Watson drove her up to dinner at Pauling's beautiful home in the hills above Pasadena. A less successful outing with Watson and Sydney Brenner took her into the honky-tonk sections of Los Angeles and Hollywood. They could see that she was deeply shocked by the girlie photos on display. She wrote as much to her parents:

> The centre of Hollywood is vulgar and characterless, and only a little less sordid than the centre of Los Angeles. Here I was shown the worst part, which is really incredibly awful. Bad patches can be found in any big city, but none of the same extent as this, even if of the same depravity.

The University of California at Berkeley, outside San Francisco, in contrast, was disappointing. Scientifically, the five-day visit was rewarding and essential for her TMV research but strangely formal, and the famed Virus Laboratory, at the top of the campus, with a view of the Golden Gate Bridge, 'the first unfriendly lab I've come to'. Instead of being invited into private homes and shown the cultural sights as she had been everywhere else, apart from one drive around the Bay, she was taken back to her hotel almost every afternoon at 5.30 and left to fend for herself until the next day. In spite of the 'perfect climate' and a quick impression that San Francisco was almost as civilised as Boston, she was glad to fly on to Washington, where she stayed in Bethesda with Alex and Jane Rich. The American capital lived up to its reputation:

> It is beautifully laid out, with vast green spaces, trees lining most streets, no skyscrapers, and a high standard of architecture. There is a large district, Georgetown, reminiscent of the best of old Hampstead, with the difference that the buildings are bright and clean, built on a hill above the Potomac river, in a good pinkish-red brick, with white plaster. Immediately outside Washington and even running nearly into the centre, also, are extensive woods in their natural state open to all.

She left for home from New York on the French liner *Liberté* a changed woman. Within the space of two months, her reputation had been considerably enhanced. The October issue of the *Scientific American* was out, with Crick's article, giving her as direct an acknowledgement as she was ever to get during her lifetime for her contribution to the great discovery. 'Watson and I,' he wrote, 'were convinced that we could get somewhere near the DNA structure by building scale models based on the X-ray pattern obtained by Wilkins, Franklin and their coworkers.' The magazine used her Photograph 51, credited to her, as an illustration. Even better, she was returning home with contacts with American virus laboratories that would form the basis of the greater part of her work for the next three years.

She left, moreover, with virus samples and promises of collaboration from leading scientists around the States, including Wendell Stanley. On board ship, she wrote a thank-you letter to Pauling, saying that she carried home with her 'the pleasantest possible memories of Cal Tech and its surroundings and its California sunshine'.

For Rosalind, Americans now fell into the category of

foreigners with whom she felt at home. So generously had she been looked after that she was able to refund the National Coal Board £200 of the £250 it had advanced towards her trip.

SIXTEEN

New Friends, New Enemies

(Winter 1954–Summer 1956)

> 'She needed a collaborator, and she didn't have one. Somebody to break the pattern of her thinking, to show her what was right in front of her, to push her up and over.'

THAT ASSESSMENT of why Rosalind missed the last two steps leading to the discovery of DNA's double helix came from the best collaborator she was ever to have. With Aaron Klug, theoretical physicist, chemist and crystallographer, Rosalind did the finest work of her life.

They met on the stairs leading up to her attic room at 21 Torrington Square early in 1954. Klug had the front room on the top floor. Rosalind's combined office and lab were at the back. Although Klug had come to Birkbeck in late 1953, they did not become well acquainted at first because Rosalind (when she was not travelling) went to lunch in the senior common room rather than with the rest of the staff in the general canteen, and also because Klug was working on a different project.

One day, however, she showed him the photographs she had taken of the tobacco mosaic virus. 'From then on,' Klug said in the autobiography written in connection with his Nobel prize in 1982, 'my fate was sealed.' The photographs were beautiful and intriguing; he offered an interpretation of some anomalies. Soon he won permission from Bernal and Harry Carlisle, professor of crystallography, to transfer to virus research. Together he and Rosalind made an ideal team, what Watson found in Crick and she had found in Mering: someone who shared and sharpened her ideas, who could argue with her and share the day-to-day excitement of discovery.

Klug, a small neat quick man with an ironic smile, was then twenty-six. South African, the son of Jewish immigrants from Lithuania, a graduate of the University of Witwatersrand with a master's degree from the University of Cape Town, he had taken a PhD from Cambridge in 1952 and come to Birkbeck the following year on a Nuffield Foundation fellowship to do X-ray analysis of biological molecules. His wife, Liebe, was a modern dancer and choreographer, and they had a small son. The couple rapidly became Rosalind's personal friends – her new Luzzatis.

Rosalind needed a new friend because she had a new enemy. She came back from America ready to write. After a year's hiatus in 1954 when she had published nothing, she had accumulated enough information through a mass of excellent X-ray photographs for a number of papers on TMV. The first was to be for *Nature* on the virus' structure and she sent a copy of the manuscript to the British virologist Norman W. Pirie.

Pirie had a proprietary interest in plant viruses. He grew his own at the Rothamsted Experimental Station in Hertfordshire and had supplied Rosalind with specimens for her research. He and his colleague, Frederick Bawden, were the Gilbert and Sullivan of British plant virology, their names linked since their landmark paper in 1936 revealed that the virus was long and thin, like a pencil or a stick of cinnamon. Watching the pair in action, the young French geneticist François Jacob saw 'old cronies who loved to play the buffoon, trading jokes and metaphysical aphorisms, all in a rapid choppy English which left me in a cold sweat'. Jim Watson, when he was working on TMV, witnessed their confident double-act dominating a meeting: 'No one could match the smooth erudition of Bawden or the assured nihilism of Pirie, who strongly disliked the notion that some 'phages have tails or that TMV is of fixed length.'

But Rosalind had found, from X-ray photographs of a kind no one had taken before, that the TMV rods were all the same length (3,000 Ångstroms). She said so in the opening line of her paper. She also said that protein subunits inside the rods were identical.

Pirie would not hear of it. Among his colleagues he was known to be difficult and abrupt with younger people, and made no attempt to conceal this, least of all from a young woman researcher with no academic rank, whose new grant from the Agricultural Research Council he and Bawden had approved. He wrote an irate letter to 'Dear Miss Franklin,' as if she were a lazy student:

> As you probably expected I have all the usual objections to this paper. But in case the faults have slipped in through inadvertence I may as well itemise some of them.
>
> You start off with a bald erroneous statement. It is true that under certain conditions of testing, the number of lesions given by a constant weight of material is greater the greater the proportion of 300 mμ particles in the prep. It is also true that preparations nearly free from 300 mμ are infective to an extent that makes it unreasonable to assign all the infectivity to one length of particle. Defined biologically therefore you cannot say TMV is 300 mμ long. Also no one has published data on any preparation in which 300 mμ particles make up more than half the total weight; so the statement isn't true physically either.
>
> This is all no doubt a pity for succinct writing but there it is.*
>
> The argument from amino acid composition to weight of the protein sub-units depends on knowledge of how many types of sub-unit there are. You, baseless, assume only one. I would be more cautious. If you want an argument in favour

* Her article, as published in *Nature* on 26 February 1955, began: 'Tobacco mosaic virus is a rod-shaped virus, of (most frequent) length 3000 Å, diameter 150 Å and molecular weight 50 million.' The parenthesis may have been added in response to Pirie's criticism.

> of such caution I would remind you of all the excitement that now accompanies the demolition of the old, baseless, tetranucleotide picture of RNA, etc. Only those who did not realise that nature should not be assumed to be simple ever swallowed the hypothesis. I suggest here that the idea of a single type of sub-unit is the least probable of all the possibilities . . .

And there was much more. Rosalind sensed that Pirie was a dangerous enemy to have. With greatest respect and thanking him for his continued help, she defended her own findings. So she should have done. She was right and he was wrong. The TMV rods *are* of the same length, and the protein subunits – as her research proved – are identical. On 7 December 1954, she replied:

> Dear Dr Pirie,
> Many thanks for your letter and criticisms, which were as you suggest, not unexpected. In particular, I was afraid that you would disapprove of the first sentence. But the E.M. photographs (Williams et al.) of the 'crystalline inclusions' breaking up do seem to show a convincing uniformity of length, indicating a fundamental unit of length 3000 Å, and the fact that such units are often aggregated end-to-end does not vitiate any of the argument.
>
> What the crystallographic data show is that the units are sufficiently similar to enable them to occupy structurally equivalent positions . . . One of the merits of the Watson and Crick DNA structure is that it explains how different bases can occupy structurally equivalent sites, and it seems that something of the same kind must happen in the virus protein. But the nature of the X-ray diagram shows that there *is* a high degree of repetition in the structure.
>
> In my paper I am now referring to the units as 'structurally equivalent'.

It is never a good idea to let your opponent know your worst fears, as Rosalind did when she concluded, 'I hope that you do not disapprove so strongly of what I have written that you will never again be willing to provide me with material to work on.'

A vain hope. Pirie was so annoyed with her for the paper that appeared in *Nature* that he stopped sending her viruses. From then on she and Klug had to grow their own. What they could not grow for themselves was Pirie's influence with the Agricultural Research Council (ARC) and its powerful secretary, Sir William Slater.

Far friendlier eyes on the same *Nature* manuscript were cast by Jim Watson. He had proposed the basic shape of TMV in a paper in *Biochimica et Biophysica Acta,* and Rosalind was building on his idea, correcting it with the help of her unique X-ray photographs. Watson wrote Rosalind from Caltech in December that he and his colleagues, Leslie Orgel and Don Caspar, had all read her draft and liked it. 'Thus the following criticisms are only of 2nd degree importance.' There followed a long list of technical points (along the lines of 'Size of subunit 5×10^2 95% = 4.5 x 10 7/37 x 3,000 over 68 = 35,000'). Watson ended, 'In spite of these criticisms, a very nice summary of TMV status . . . We are naturally curious as to what you are doing.'

Whatever impression Watson's flopping shirt-tails and lace-less canvas shoes had made at Cambridge, there was nothing untidy about his science. His notes to Rosalind were in tiny writing as precise as a typewriter's, with clear diagrams and formulae: the marks of a man who saves neatness for where it counts.

Rosalind's second TMV appearance in *Nature* followed a few months later in a joint paper written with Barry Commoner of Washington University in St Louis, who had sent her a preparation of his abnormal protein called B8 associated with tobacco mosaic virus after their meeting in St Louis the previous summer. Commoner (later the well-known ecologist) had the advantage of equipment she lacked. 'There not being an ultracentrifuge anywhere in this college,' she wrote him, 'I concentrated the solution you sent me simply by allowing it to evaporate slowly through the dialysis bag.'

Frequently she went to Cambridge to talk with Francis Crick, who had returned from Brooklyn. Like Watson, he was deep into RNA and virus structure. For Rosalind, Crick had moved into the 'brilliant' class that stirred her hero-worshipping tendency. In him she saw the gravitas, demeanour and acute reasoning she expected of a great scientist. 'She would do nothing without clearing it with Francis,' Klug observed. Yet Rosalind could scold even Crick, and was heard lecturing him, 'Facts are facts, Francis.'

(When another of Rosalind's heroes, Linus Pauling, was awarded the Nobel prize for chemistry in late 1954, Rosalind was horrified that the BBC news report concentrated on Pauling's politics rather than his achievements as the greatest chemist of the time. She got her Birkbeck colleague Wolfie Traub to call someone he knew at the BBC and ask them to rewrite the bulletin.)

She and Klug were hard at work on two joint TMV papers. Between them they agreed that Rosalind would concentrate on the rod-shaped viruses such as TMV, while Klug would move to the spherical viruses affecting other plants. (Plant viruses appear as rods or spheres.) They needed assistance, and with advertisements in *Nature* struck gold. Two future Fellows of the Royal Society joined them as research assistants, registered for a PhD and threw themselves into the work. John Finch joined in January 1955, from King's College London (where he had heard vaguely of some difficulty between Rosalind and Maurice Wilkins, but nothing more). Kenneth Holmes, from St John's College, Cambridge, joined in July. Rosalind decided the division of labour: Holmes would work with her, taking rod-shaped viruses as his thesis topic; Finch would tackle spherical viruses with Klug, with Rosalind in overall charge.

Holmes and Finch saw at once that Klug and Franklin had harmonising skills and temperaments. Both could be aggressive when science was at stake – Klug so much that Rosalind had to restrain him. 'Don't bully, Aaron,' she would chide. Klug, for his part, a theoretician, in no way saw Rosalind as a mere experimentalist, an unequal partner. 'It takes imagination and intellect to know precisely what experiments to do, to design them, prepare the specimens and then to observe the results,' he said retrospectively. 'She worked beautifully. Her single-mindedness made her a first-class experimentalist, with the sort of skill that blends intelligence and determination.' It was not, he emphasised, 'just good needlework'.

Holmes, who worked closely with Rosalind, got to know her well, and before long, in his words, 'would have gone through fire and water for her. She *was* prickly and difficult, especially at first, unable to put people at their ease. There was a forcefulness about her manner . . . a barrier to be overcome if you wanted to get to know her.' His fiancée, Mary Scurby, who worked in the Birkbeck research library, found Rosalind hard to talk to. She was surprised to learn Rosalind had much admired the circular red felt skirt she had worn to a concert to which Bernal invited them all as a staff outing. John Finch did not work as closely with Rosalind and remained a bit wary of her, but admired how she nursed her viruses as they grew in their gels; he saw fine work, beautifully done.

With Holmes, as with the others, Rosalind was very insistent that the work be done the way she wanted it done. One day, very proud of himself, Holmes cured a problem with the Beaudouin X-ray camera she had brought from France. Cleaning its tubes to get a vacuum was difficult, and after struggling, he devised a vacuum gauge to be fitted to the X-ray tube. To Klug's and Finch's amusement, Rosalind was not impressed. 'We want to make X-rays, not measure vacuum!' she scolded Holmes.

Holmes thought she was beautiful. Looking back, he believed that, of course, Bernal would have made a pass at her – 'a woman that attractive' – if he had dared. It was an office sport to watch, when there was a new secretary, how Bernal would keep finding excuses to come into the office. But Rosalind's formidable facade held off even this most intrepid of womanisers. She knew his weakness, however. Everybody who worked in the Torrington buildings noticed the creaking floorboards from Bernal's flat and the strange women coming downstairs in the morning. There was an embarrassing incident when Bernal's car was stolen from a street in Soho. Vittorio Luzzati was visiting at the time (Bernal had great respect for Luzzati's abilities at X-ray analysis), and Rosalind tried to explain to him why Bernal's car should have been parked in London's red-light district. 'C'est un endroit où sont les femmes!' she said. 'She didn't know the French word *putain* [prostitute],' Luzzati recalled. 'She was very puritanical.'

With Klug, Finch and Holmes, and another research assistant, James Watt, subsidised by the National Coal Board, who worked with her on coal, she had a quasi-family. She proudly added to her curriculum vitae 'Leader of "ARC Group" at Birkbeck'. One of their unofficial functions was to act as buffers in her relationships with the rest of Birkbeck.

Buffers were needed. When Rosalind's technician, Bryon Wilson, told her he intended to take advantage of Birkbeck's scheme of offering staff a week off for education – he wanted to work towards an O-level certificate – she questioned his intellectual abilities in words that stung. When he persisted, she transferred him to her coal research assistant and ignored him after that. (Wilson eventually got a doctorate and became a lecturer in microbiology at St Mary's Hospital Medical School.)

The same qualities of abruptness, reserve and suspicion that had been noticed at King's College were noticed at Birkbeck too. Bernal's people saw themselves as a family into which Rosalind didn't fit. According to Wilson, 'Everybody called her "Rosy" behind her back. She didn't like it!' She tended to pass people on the staircase and not say hello. She did not come to tea in the lab as the rest did. Holmes, one of her great admirers, acknowledged, 'She was not a great communicator.' He attributed it to her shyness. The clerk to the college, A.J. Caraffi, on the other hand, blamed her lack of academic status. This left her, Caraffi thought (although personally he saw her as a woman of wit and sparkle), with a feeling of 'not quite belonging' to the staff.

She was notorious as a non-conversationalist. At lunch in the faculty dining room (where she had at first trouble gaining admittance because she was considered non-faculty or 'external staff'), one day she plucked up courage to comment, 'It's a fine year for mushrooms,' and lapsed into silence again.

She had an outright clash with Stan Lenton, the college's much-admired steward in charge of technical services. Lenton grew very irritated with the curt way in which Rosalind would ask for new equipment. Things came to a head when she ordered a massive transformer from Paris. The monster gadget turned up at a Customs freight department in south London. From long experience, Lenton knew the forms and procedures necessary to get things through Customs without being charged duty, but Rosalind declared she was 'quite capable' of dealing with it herself. A week later she knocked on his door and acknowledged defeat. He solved the problem for her and, 'We never looked back after that.'

One reason to hold herself aloof was that so many of Bernal's staff were extremely left-wing or even outright members of the Communist Party. The social divide between her and them was real. People at Birkbeck were aware that there was money behind her. As someone put it, 'She lived in Kensington and she travelled.' One evening a big car drove up to 21 Torrington Square; from the back a woman in evening dress asked for Rosalind. The porter who went up to call her found her in her lab coat and was told she would be down in ten minutes. In exactly ten minutes, she emerged, elegant, transformed, and was driven away.

If Rosalind made much the same impression at Birkbeck as at King's, the great difference was that at Birkbeck, she was leader of a team doing superb work; she was neither isolated nor unappreciated nor unprotected. Bernal thought the world of her and saw that her firmness inspired those around her to reach the same high standard.

* * *

If she had lacked a collaborator at King's, she now had three, and in mid-1955 acquired a fourth. The American biophysicist Don Caspar, whom Watson knew at Caltech, held a PhD from Yale, with a thesis on TMV. In mid-1955 he left Caltech to spend a year at Cambridge as a postdoctoral fellow in molecular biology, and asked Rosalind (with many friendly references to 'Jim') if he might work with her for a time. Rosalind replied that she was interested but had no funds and also had two new people to look after, but 'if that doesn't put you off we shall be very glad to have you'.

Caspar was an ebullient twenty-seven. Having known Rosalind only through correspondence, when he finally met her in the summer of 1955, he was taken aback. He had expected a dour English bluestocking, 'and she turned out to be an attractive vivacious young woman'. The ingredients for compatibility were there: Caspar was foreign, very bright, with a Jewish background and an intense preoccupation with their mutual special subject. He was unmarried and had a disarming friendliness. One day when he arrived to see her, Rosalind was outside on the steps, with Ken Holmes and James Watt. When the small moustached smiling figure hailed her, 'Hey, Ros!' her companions waited for the explosion. There was none. Somehow the young American got away with it. 'I had no hang-ups with her at all,' Caspar said, and was astonished that others did.

Caspar's work and Rosalind's meshed perfectly. At Yale he had discovered, as she had heard from Jim Watson when she was at Caltech, that the tobacco mosaic virus has a hole straight through the middle. The hole came as a surprise because the virus's centre had been thought to be full of the nucleic acid, RNA.

Rosalind was on her way to an answer to the related question: where is the RNA, if not in the central cavity? From her X-ray pictures which gave her better data than had been available to Caspar and Watson, she discovered that the RNA is deeply embedded between the protein subunits, winding around its

Summary of the first analysis of the structure of TMV shows the protein subunits ranged around the hole at the core, with the RNA embedded between the protein subunits, winding around its inner groove like a twisting thread. This diagram was published by Klug and Caspar in 1960.

inner groove like a twisting thread. She found also that the protein surface of the TMV is like a knobbly screw.

Caspar was employing, just as Rosalind, Klug and Holmes had in their own research, the very new technique, developed by Max Perutz, of isomorphous replacement, or heavy atom substitution. That is, they introduced some atoms of a heavy element such as mercury or lead into the virus protein. The difference between the X-ray patterns produced by crystals of the molecules with and without the heavy atoms then reveals the structure. Revealingly, the graphs plotted by Caspar and Rosalind, when superimposed, showed the exact distance of the RNA from the centre of the virus.

This information was of more than abstract interest. Once the internal configuration of the virus was known, the way it works could be understood. The protein surrounds and isolates the nucleic acid (RNA) until the virus has infected a host cell. Once inside the cell, the RNA is released from the protein and begins generating new virus – the dreaded infective process.

Rosalind's work was fascinating but isolating. No one outside the field could understand it. When she gave her parents two simple books on viruses, she was disappointed when they said they couldn't understand them. Relief from the intense concentration on work came in various ways. In London she enjoyed the theatre and the cinema, and was a member of the National Film Theatre. As ever, she threw herself into her holidays abroad. In the summer of 1955 she made a return visit to Yugoslavia for a month, and then went on a cycling trip in Normandy with Anne Piper.

Wherever she was, she found tremendous pleasure in the children of her friends and relations. She was godmother to one of the children of her cousin Catherine (Joseph) Carr and regularly went for weekend visits to the Carr home on the Sussex Downs. Catherine had not only married 'out' but further rebelled against her family by choosing a working-class man. Rosalind got along very well with her husband. She loved playing with the children and bringing them presents. To Catherine, her cousin was a single woman who should have married and had children of her own but who was too bright for most of the men she knew.

Another former school friend, Jean Kerlogue, often stayed at Rosalind's flat with her family. She was, like so many others, astonished by Rosalind's amazing ability to produce exactly the right toys and games for the children. Yet for Rosalind choosing good presents, like cooking well, was more than a matter of flair; she did it with the same kind of astute observation and eye for detail that she brought to her experiments.

Friends tried to introduce her to men. Evi Ellis, who had returned to London from Chicago, was married and living in Notting Hill. She arranged a dinner party to bring together

Rosalind and Ralph Milliband of the London School of Economics. Milliband was impressed by Rosalind but nothing came of it. No wonder, perhaps, considering her pervasive unapproachability. Rosalind told Evi that a man who had a flat on the same floor of Donovan Court had asked her, going up in the lift one night, if she would like to come in for a drink. 'She didn't seem to know why he had asked her,' Evi said.

Various of her married friends have reported hearing Rosalind maintain that women should not have both a career and children because it was unfair to the children. As the 1950s progressed, however, Rosalind saw among her close friends women who managed both very well. Anne Piper had four children and was an executive with the John Lewis Partnership; Mair Livingstone and Denise Luzzati were both doctors and mothers of two. Hearsay apart, however, there is no evidence that Rosalind ever made a deliberate decision to be a scientist rather than a wife. What seems more likely is that she was afraid of intimacy and could find it only in professional relationships or with married couples. Colin Franklin marvelled at Liebe Klug's tolerance of the closeness between Rosalind and Aaron. But it was no mystery to the Klugs, who long remembered the mechanical fish she brought from America for their small son. 'She was a good aunt,' said Klug.

Once they had deciphered the structure of the tobacco mosaic virus, Rosalind and her team were ready to construct a model. As components, they needed, not the shining metal plates that stood for atoms in the Watson-Crick model of DNA, but something with a longer shape to represent the protuberant subunits of TMV. Bryon Wilson (at that point still working with Rosalind) thought that a bicycle handlebar grip might be about the right shape. As Birkbeck was not far from Oxford Street, he went down to Woolworth's and asked for some handlebar grips. How many? When Wilson replied 'two gross [288]', the salesclerk called the manager. It could only be a lunatic who would ask for the shop's entire stock. Wilson got them in the end, the model, and many subsequent versions, was constructed and was much in demand for exhibitions.

Science in the days before the computer and the photocopier involved many tedious jobs. Another of Wilson's was to go to the mathematics library in Bedford Square and copy down by hand Bessel Function tables. The only handbook available in Birkbeck was in the library and could not be taken away, and Rosalind could hardly be expected to perform such drudgery. In the same spirit, the ARC acceded to Bernal's request for Rosalind to hire a 'computer' – that is, someone to perform computations. It had no wish to have 'Miss Franklin's time occupied on computations that could be done, as you suggest, by a girl of eighteen'. As the official rate for an 18-year-old girl in the 'Machine Assistant' grade was five pounds a week, the ARC agreed, and a 'girl' – a Mrs Cratchby – was hired at £350 a year.

Before long, Rosalind's group was working on a whole minestrone of plant viruses – potato, turnip, tomato and pea. Raiding the Birkbeck fridge, Caspar and Klug found crystals left from the 1930s by Bernal and Carlisle. Caspar took the Tomato Bushy Stunt and Klug the Turnip Yellow Mosaic. Rosalind, meanwhile, was scouring the world for virus samples. She asked a virologist at the University of Wisconsin for ten milligrams of Pea Streak. The most convenient form in which to send it, she recommended, was 'a rather concentrated solution or an ultracentrifuge pellet'. She added that 'Jim Watson (of the DNA model) is back in Cambridge, and is also interested in these things, and between us we want to look at as many viruses as possible.'

Watson was indeed back at the Cavendish, to spend the second half of 1955. He had a year's leave before starting his new post at Harvard as assistant professor of biology so that he might work with Crick on principles of viral structure. As Watson had shown an interest in the potato virus, Crick suggested that Rosalind let Jim know if she were intending to work on this herself. If so, he said (in a cooperative spirit not in evidence two years earlier over DNA), Jim would leave that virus to her to avoid unnecessary duplication.

Her team was complete, the work was going beautifully, the results were plentiful. But how long could they all stay together? The threat to her group began almost as soon as Rosalind realised how good it was. When Bernal saw Norman Pirie, just as the Agricultural Research Council replaced the Turner and Newall Fellowship as Rosalind's paymaster, Pirie was still fuming over Rosalind's paper on TMV's fixed length. Pirie's bad temper amused Bernal, who knew all the same that Bawden and Pirie were very influential with the ARC.

The ARC grant was at most for three years – that is, until the end of 1957, and every item of new equipment and staff had to be haggled over. Before going on holiday, Rosalind made a list of her requirements: a centrifuge, an X-ray tube, a Geiger counter spectrometer, a biochemist, two research assistants and more working space. She also asked to be upgraded from Senior to Principal Scientific Investigator and to have Dr Klug's future planned for. Klug's Nuffield grant would run out in a year and he had to leave his flat in a few months. He could not move without knowing whether the ARC would allow Rosalind to use its funds to employ him.

The early omens were not good. Jim Watson tried to help by speaking to Lord Rothschild, chairman of the ARC, who lived in Cambridge. 'I've had a long talk with Victor Rothschild about your ARC grant,' Watson wrote Rosalind. 'His reaction was very sympathetic, indicating that he would write Slater immediately.' Then Watson added, 'Perhaps it would be a good idea if we could talk again before anything positive is done so that the mistake of applying for too little could be avoided.'

Rosalind was bitter that those like herself who did research full-time had lower academic rank than those who taught as well. Her salary symbolised her discontent. She was still simmering about having been cut back to £1,080 by the Agricultural Research Council.

In mid-1955, after university salary scales were revised with a view to raising women teachers to parity with men over a period of seven years, Rosalind was entitled to an increase of £150. The college asked that her pay be raised to a total of £1,250. The ARC once again refused, on the grounds of her age.

She was furious. To Bernal, who she knew would understand, she set out her case:

> My age is 35, I have been doing full-time research continuously for the past 14 years, and obtained my PhD 10 years ago. I cannot believe that there is any rule which prevents the ARC from paying a salary greater than £1,080 p.a. to a person of my age and experience. My present salary is less than I should be receiving if I had made a career either in university teaching or in the Scientific Civil Service, and is also less than the *average* received by physicists of my age. In view of the fact that I have no security of employment, and nevertheless hold a position of considerable responsibility, this seems to me entirely unjust.

There was unquestionably an element of choice in Rosalind's anomalous position. She did not wish to teach, nor to move to a research institution. But her weakness with the Birkbeck hierarchy reflected Bernal's own. He was at permanent loggerheads with the Master of Birkbeck. To keep his department's science in the top rank, he depended (a bit like John Randall) on scraping up grants from a wide variety of sources, even the Flour Millers' Association. The ARC, moreover, as donor agencies went, was both mean in spirit and unaccustomed to doling out its money to women. The shaving off of the paltry twenty pounds from Rosalind's deserved total was just as insulting as she took it to be. By the end of 1955, the ARC gave it to her, bringing her annual salary to £1,200 – still lower than her entitlement.

Rosalind often presented the face of an embattled woman to the world but she had a great deal to be embattled about. When she went to see Slater at the ARC at the end of September, fortified with Bernal's warm endorsement for her group's rapid progress, she got a stony reception. Slater refused her every single thing she asked for. She needed an X-ray tube? There was one in Sheffield. The only way her work could continue to be financed would be to move it to Cambridge – and only one person could go. Even there, he wanted an absolute halt to the use of virus samples – from Berkeley, Tübingen or anywhere else abroad: he strongly disapproved of what he called 'second-hand material'. (This ludicrous and provincial diktat was, in its way, as unethical as Randall's telling Rosalind in 1953 to stop thinking about DNA. Science, of all intellectual disciplines, knows no geographical boundaries.)

On her own position, moreover, Slater would not consider raising her to the grade of 'Principal Scientific Investigator'. When she said she ought to have had that title four years before, he disagreed; to reach that at thirty-one, he said, it would be only for the 'exceptionally distinguished'. Calling her 'my dear', he treated her like a little girl and told her that she ought to work under the constant guidance of a biochemist because otherwise she could not hope to understand the biological side of her work.

Rosalind argued back, point by point. Heated rebuttal from a woman was not what Slater welcomed. He refused to discuss Klug's future. He was badly upset by her suggestion that the ARC provide lab space elsewhere in the University of London: that was the university's business. She left in tears, and telephoned Klug in distress from Piccadilly Underground station.

Making careful notes of this bitter meeting in order to re-present her requests in writing later in the year, Rosalind commented to herself, 'Presumably somebody has protested my work with Commoner.' She would not have had to look far for a suspect. Norman Pirie had many reasons to object to her work quite apart from disagreeing with her TMV measurements. Because of his left-wing views, he could not travel to the United States and was cut off from friendly collaboration with American virologists. He had particularly cool relations with one of Rosalind's most helpful contacts, Wendell Stanley of the Virus Laboratory at Berkeley. Stanley, unlike Pirie, had won a Nobel prize for virus research, and had given Rosalind virus samples and was credited in her published papers.

Defiant in the face of Slater's blatant parochialism, Rosalind put in her progress report for the year 1955 that her group was in close touch with the virus-structure studies being carried out 'in the laboratories of Berkeley (USA) and Tübingen (Germany)' who 'regularly send us their new preparations for study, and our results are closely interconnected with theirs'. Until her group could make its own preparations, she said, 'we must remain dependent on the generosity of research workers overseas'.

She stressed that just as important as the new centrifuge *urgently* needed was an agreement to put Dr Klug on the ARC payroll: 'his position is such that he must make decisions very shortly about other future possibilities. It is therefore important that a decision be made.'

Rosalind had tied her future to Aaron Klug's. His friend, the writer Dan Jacobson, observed the relationship. 'Meeting Rosalind was *crucial* for Aaron,' said Jacobson, who had come to London from South Africa (via Israel) about the same time in the early 1950s. 'It was the turning point in Aaron's career. He was dissatisfied at Cambridge, could not see the way ahead and admired Bernal despite his politics. And Rosalind was *fascinated* by Aaron. It was a great bond that they were both Jewish – and not the same kind of Jew. They thought the same way – but differently, rather, they understood each other's minds.'

Jacobson, expanding on his theme, said: 'It is not too strong to say that Rosalind *loved* Aaron. Not in a sexual way. It was a meeting of minds – and of two very different kinds of mind. Aaron was imaginative and playful and artistic as well as a superb scientist. He was an expert on movie Westerns and wrote a skit with me. She scolded Aaron and told him that play-writing was a waste of time for a scientist.'

Part of Rosalind's fascination was for the Klugs' bohemian style of life. Klug felt she was shocked by the way they were living – at the top of a shabby five-storey Victorian house north of Regent's Park, washing up in the kitchen in front of guests, with friends from South Africa staying in the attic. When she was there, she would sit with her straight back on the edge of her chair. She never, as the others did, sat on the floor.

Grant worries apart, Rosalind was at the top of her profession and enjoying it. Invitations poured in, and her contacts were wide. She held an impromptu tea party at her flat on the occasion of a conference on semi- and non-crystalline materials held by the Institute of Physics on 18 November 1955. Jacques Mering was there from Paris; also Drago Grdenič from the University of Zagreb (whom she advised to look up her mathematician friend, Simon Altmann, at Oxford) and Alan Mackay, a bright spark from Birkbeck Crystallography.

In 1956 she was off on another round of conferences, none of which slowed the steady flow of papers from her lab in various combinations of authorship – Franklin and Klug, Klug and Holmes, Klug and Finch, Franklin and Holmes, Franklin, Klug and Holmes. Even when in London she strove to keep herself informed as widely as possible. Fred Dainton, her former Cambridge supervisor, was touched when she turned up to his lecture on a subject utterly unrelated to her own research. 'She seemed to me much more womanly, much less prickly,' he said, even capable of 'a little gentle teasing'.

At many scientific meetings, it goes without saying, she was the only woman on the programme. Most of the women present were wives, expected to go shopping or sightseeing during the formal sessions and to join the men only for social gatherings. At an important invitation-only conference held by the Ciba Foundation in London in March 1956 on 'The Biophysics and

Biochemistry of Viruses', the printed programme advised: 'The wives of overseas members (but not those of British members) are invited to join the group for lunch.'

Rosalind stood out on the otherwise all-male select guest list of thirty-four, which included Watson, Crick, Wilkins, Bawden, Pirie, Caspar, Klug, and Robley Williams of the Virus Laboratory at Berkeley. She delivered, on behalf of her co-authors Klug and Holmes, a paper on 'X-ray Diffraction studies of the Structure and Morphology of Tobacco Mosaic Virus'. Williams, who was at Cambridge on sabbatical, led off the discussion on the question Rosalind's paper had raised: whether there was just one RNA molecule in the TMV virus or many. Watson and Crick came in on Rosalind's side and Pirie threw sharp and witty barbs.

A few weeks later much of the same cast reassembled in Madrid. Until then Rosalind had avoided Spain, out of hostility to Franco, but science overrode politics. With an International Union of Crystallography symposium on 'Structures on a scale between the atomic and microscopic dimensions', with social receptions in the Plaza de la Villa and the Hotel Wellington, with her name listed once more among the top in the field, and the promise of a week of excursion afterwards, she could hardly wait. She wrote to Adrienne Weill, 'As I have never been there, I would be happy to go anywhere at all – in the cities, to the seaside, to the caves at Altamira, etc., etc.'

The symposium was a jolly gathering of professional intimates. A photograph shows an assured Rosalind in ideal circumstances, flanked by her three favourite scientists – Crick, Caspar, and Klug – as well as Odile Crick and John Kendrew, in an exciting new foreign venue. When the meeting was over, Rosalind decided to go with Francis and Odile Crick on a tour of southern Spain. She tried to get Klug to come along, but failed. Having read Hemingway, Klug was determined – in the face of Rosalind's strong disapproval – to remain in Madrid to go to the bullfight. So in a threesome once again, she went to Toledo, Seville and Cordoba. She got on very well with Odile Crick, but did not indulge in any feminine confidences and they never conversed in French. The couple found her great fun. They knew nothing about her private life. 'But,' said Crick, 'did anyone?'

She and Don Caspar had two papers coming out in *Nature*: not joint, but companion publications. In fact, she had written every word of his so that her own could appear. Caspar, like many scientists, found writing up results a tedious chore compared with the fun of experiment. Together their papers proclaimed the news: that the tobacco virus is hollow (his paper) and that along the groove between protein subunits, the RNA winds like an inner lining (her paper). Their respective graphs, when superimposed, nicely revealed the exact location of the RNA: at a radius of about 45 Ångstrom from the axis of the intact virus.

She was still fighting for the future of her group. Even before she left for Madrid, she wrote what was, as scientific correspondence goes, a passionate letter to Slater, telling him that the work her group was doing 'is concerned with what is probably the most fundamental of all questions concerning the mechanism of living processes, namely the relationship between protein and nucleic acid in the living cell . . . The plant viruses consist of ribonucleic acid and proteins, and provide the ideal system for the study of the in vivo structure of both ribonucleic acid and protein and of the structural relationship of the one to the other.' There was no doubt, she said, that the results obtained would throw light also onto the problem of animal viruses (such as polio). 'Moreover, in no other laboratory, either in this country or elsewhere, is any comparable work on virus structure being undertaken.'

Rosalind's work was yielding a rich harvest of papers. By mid-July 1956 she had sent a paper to *Biochimica et Biophysica Acta* comparing three strains of TMV with a cucumber virus. For the same journal, she and Holmes did another paper giving precise measurements about the internal shape of the cucumber virus. And the papers by herself and Caspar that had come out in

Nature in May were recognised as technically remarkable for demonstrating the application of the new technique of isomorphous replacement to a non-crystalline specimen.

More cheering news was that another Gordon Conference beckoned from New Hampshire. She was being invited to one on nucleic acids this time; Don Caspar was going too. Best of all was the glimmer of hope about future funding. Robley Williams, among others, had suggested to her that the ARC was not her only recourse. Why did she not consider applying to the US National Institutes of Health in Washington for funds? She did not take much persuading. The American connection began to look like a lifeline.

SEVENTEEN

Postponed Departure

(May–October 1956)

THINGS WERE GOING WELL for the extended family in 1956. Ellis Franklin and his sons David and Roland were directors of Keyser Ullmann, a new firm formed by merger of Keyser's with the banking partnership of Ullmann & Co., Colin was an editor with Routledge and Kegan Paul. There were many grandchildren. Jenifer was running the Goldsbrough Orchestra and working also with the Architects Co-Partnership. Aunt Mamie – Mrs Helen Bentwich – was chairman of the London County Council. Uncle Herbert – Lord Samuel of Toxteth – was the spokesman for Anglo-Jewry, replying to the toast given by the Duke of Edinburgh at a banquet at the Guildhall to celebrate the tercentenary of the Cromwellian settlement of the Jews in England. Rosalind, director of the virus project at Birkbeck College, was off on another invited tour of the United States.

There was no grubbing around for travel money for her second American trip. The Rockefeller Foundation offered to pay her expenses to the Gordon nucleic acid conference and to tour American laboratories afterwards. Before leaving for the United States in June, she had a medical check-up with Dr Linken of the University College London's Student Health Association; this was a routine examination required of those exposed to radiation risks in their work. No problems were found. Her worst irritant was the failure of a Newcastle supplier to send the promised representative to regulate her Birkbeck group's microphotometer.

She wrote a stern letter saying she would shortly be leaving the country for several months and her assistants needed to be able to carry on their work in her absence.

The generous travel grant from the Rockefeller Foundation did not relieve her dependency on the Agricultural Research Council. From Birkbeck Bernal wrote Sir William Slater, the irascible secretary of the ARC, to ask permission for Rosalind to accept the Rockefeller money and to be away for two months visiting virus labs in New York, St Louis, Madison, Wisconsin and Berkeley, California. In a remarkably deferential tone for an egalitarian Communist, Bernal explained to Slater:

> Her work during the past two years has been mainly concerned with materials prepared in these laboratories, and it would be of great value to her work here if she were able to renew her contacts with these laboratories, become better acquainted with the extensive progress made in virus research during the past two years, and make preparations for extending her structural studies to new viruses and virus derivatives. This grant is naturally subject to approval by the Agricultural Research Council of Miss Franklin's absence, but as it is evidently in the interests of the advancement of her research I should imagine there would be no objection.
>
> I am accordingly writing to ask your permission for Miss Franklin to be absent for two months, from mid-June to mid-August.

Slater replied through a deputy that the council had no objection to Miss Franklin accepting a travelling grant for the purposes described and noted her intended absence for two months.

One of life's unsung pleasures is the comfortable return to a place once forbiddingly strange. Embarking on her second American trip, Rosalind knew that she must not over-plan her schedule as she had in 1954, but rather to leave room for unexpected invitations. She was prepared for: good weather, good science, healthy outdoor pursuits and warm hospitality. She was primed to spot the differences from her last visit.

The hopscotch transatlantic flight was easier than before, stopping just at Shannon and Boston in order to reach New York. Hyper-sensitive to light, she brought with her an eye mask to help her attempts to sleep. On arrival she noticed the first change. In the immigration lounge at Boston, there was 'a table with a copious supply of coffee, for one to help oneself. Such details make a lot of difference to a journey – 5 cups so far this morning.' From New York she went immediately to Baltimore, to attend an important conference at Johns Hopkins University on the chemical basis of heredity. There she welcomed 'an abnormal patch of "cold" weather, well down in the seventies' as well as the array of great names. Being driven back to New York at night, she was surprised, arriving at 2.30 a.m., to find the streets full of life and some shops open.

With Anne and David Sayre in New York, she went on a food-shopping expedition to an impressive European-style market 'but on a American scale with wonderful and varied foods of all kinds'. Her first impression, she told her parents, was that the standard of living was even higher than two years before 'by quite a large amount. All luxuries seem to be commonplace and it is hard to know how to begin to return the lavish hospitality I receive.'

Then it was back to the familiar grounds of the New Hampton School in New Hampshire. She wrote every few days to Aaron Klug; at last she had someone with whom she could share the scientific information and gossip she was picking up. Still worrying about the financial future of her Birkbeck virus group and of Klug in particular, she had learned that if Klug wished a year in the United States, Don Caspar might be able to help him get to Yale. Also, that Francis Crick would try to find room for him at Cambridge. She had forgotten, she told Klug, to discuss with him in advance how much she ought to tell the scientists she met about their recent results at Birkbeck. 'However, as it's better to tell too much than too little I told Jim [Watson] about the microsomes [a small complex within a cell]. He seemed a bit surprised that we'd been able to get them . . . Jim intends to work entirely in microsomes.'

In New Hampshire, she was prepared to enjoy herself and did. Afternoons were free for walking and swimming. Women scientists were no rarity; there were a good many at the Gordon Conference. One of these, Helga Boedtker (Mrs Paul Doty), a Harvard biochemist, much enjoyed talking with Rosalind and was very impressed with her total dedication to science.

The conference subject – proteins and nucleic acids – was far more interesting than that in 1954 on coal and by now there were many familiar faces, among them Barry Commoner, Erwin Chargaff, Alex Rich, George Gamow, Heinz Fraenkel-Conrat, and Wendell Stanley. Rosalind was now a recognised member of an international fraternity of clever people who talked each other's language, knew each other's work and crisscrossed the country and the Atlantic. It was fun seeing the same faces one week in Baltimore, the next in New Hampton or Woods Hole and shortly after in San Francisco or Madrid.

Don Caspar was there too. She had already seen him at Baltimore and they got on extraordinarily well. Rosalind's well-guarded heart was beginning to acknowledge a compatibility that went beyond companion papers in *Nature*. Together they arranged that Caspar would come out to Berkeley to work with her while she was there; also, that she would visit him at his home in Colorado Springs later that summer. In her letters home Caspar's name was censored out: he was 'a friend from the East'. To Klug, however, she reported, 'Don is injecting rabbits now!'

Caspar was both susceptible to women and inexperienced with them – in his view, 'because of my strong-willed mother'. He knew he always felt more comfortable with women than men and that he had 'more rapport with Rosalind than with most of the men I knew'.

When the Gordon gang lined up for the group photograph, however, it was not Caspar who had pride of place beside Rosalind in the front row, but Maurice Wilkins. The two sat untouching and unsmiling as the others beamed for the camera.

While she was in New Hampshire, from the Royal Institution in London came an invitation that could scarcely have pleased her more. Sir Lawrence Bragg asked for a model of her work on the helical and spherical viruses to be exhibited in the International Science Hall of the Brussels World's Fair, scheduled to open the following April. His young staff, Bragg said, had been very impressed by her recent models when he showed them in his recent lectures. She would have financial help for construction and was to think of something on a striking scale, with simple explanations for the general public.

Going back to Woods Hole on Cape Cod, with familiar faces and without a hurricane, was like a homecoming. She recognised the place as a happy mixture of serious science and seaside resort (the small town holds both the Marine Biological Laboratory and the Woods Hole Oceanographic Institute). She told her mother, 'The beach is crowded but exclusively with scientists and their families, and the village and surrounding country are really remarkably unspoilt. Woods Hole itself has a small fishing port which might be anywhere in Europe except that it's somehow more efficient and huge refrigerated trucks take the fish straight from the boats to New York (about 250 miles) and Boston.' Writing while sitting on the beach, she observed:

> The quality and quantity of everything in the shops is astonishing, and prices are not high even at the official rate. Food and lodging costs more than at home but apart from this the cost of living is high only because standards are high.
>
> The sun is hot, and I must get into the water.
>
> Love from Rosalind.

From there she went on to her engagements across the continent, then flew to Los Angeles. She had set aside four weeks for the West. UCLA (the University of California at Los Angeles, she reminded her parents) was sheer fun. Sam Wildman, professor of botany, gave a party at his home in Santa Monica where, over martinis, he gave Rosalind a lot of ribbing for being British, and did an imitation of an Englishman saying, 'Eh wot?' Rosalind (whom Wildman had found 'a pushy woman' in her earlier visit to his lab) took it in good humour, then got her own back. She told the story of the Englishwoman who arrived in America to study American culture. How long would it take?, the visitor asked. 'A week,' was the reply, 'or maybe two.'

They all had a good laugh. William Ginoza, who worked on TMV, found her 'a very sweet person, very attractive and very ladylike, I was really taken by her'.

Wildman's modern house deeply impressed Rosalind. On a hilltop with views on three sides and behind a lovely garden, it looked 'like Kew Gardens with everything grown too big'. The house, she told her parents, had been 'built for its present owner' (something almost unheard of in England at the time); also, wonder of wonders, a spare bedroom 'with its own bathroom'. Even better:

> An important feature of this and similar houses in this part of the country is a well-equipped workshop. All the people I meet seem to make nearly all their own furniture. It is a curious result of high standards of living – *everybody* earns a lot, so nobody can afford to pay anybody else. Private workshops are well-equipped with electric saws, etc., and the furniture is beautifully made.
>
> It's a healthier form of competition with one's neighbours than the size of cars and refrigerators – which seem to be people's chief concern in the East. In fact, this part of the country retains quite a few of the features of very new country – considerable pioneering spirit and intense local pride.

Rosalind disapproved, however, of 'another and rotten new form of inter-neighbour competition in large families'. She attributed the baby boom to the propaganda of baby food and gadget manufacturers: an 'astonishing illustration of the power of commercial interests'. However, staying with one expectant family, the Albert Siegels, where she slept on the sofa, they talked about the differences between American and British education. Rosalind threw at them a question from the English 11-plus examination: 'Supply the next three letters in the series OTTFFSS.' It confounded them all.*

Games apart, their discussions were wide-ranging and the Siegels were impressed with what Al Siegel called her 'deep humanity'.

Was this sweet, laughing, humane guest the same woman who was so aloof in the Birkbeck corridors, curt with the staff, silent at the lunch table and belligerent with the Agricultural Research Council? Memories of Rosalind treasured by those who met her abroad are, like holiday snapshots, bright and smiling, while many of those from England are dark and dour. The American vignettes are perhaps coloured by hindsight, also by a wish to erase the ugly 'Rosy' of Watson's *The Double Helix*. To William Ginoza, 'She was not the way Jim Watson said.' If, however, there were indeed two Rosalinds, Americans got the sunny side.

After UCLA, she was driven the twenty-six miles east to Pasadena to give a seminar at Caltech. She was rewarded with the mountain trip of her dreams. At Caltech 'a very brilliant ex-Italian virus man' (whom she did not identify to her mother but appears from correspondence to have been Renato Dulbecco) asked her if she preferred mountains or the seaside. Controlling her eagerness, she replied 'mountains', whereupon he asked if she would like to go camping. Would she mind walking? And carrying sleeping bags as well?

* The answer is ENT – on the basis that 'O' stands for one, 'TT' for two and three, 'FF' for four and five, etc.

Would she mind?! With three others she set off at 6 a.m. on a Friday. By 11 a.m.

> we had driven 220 miles to a point 8,500 feet high at the foot of Mt. Whitney (14,495 ft), the highest peak in the USA (outside Alaska), and carrying sleeping bags, blankets, and food for 24 hrs. It's no good trying to describe mountains they all sound the same, but this was incredibly beautiful, and got consistently and amazingly more beautiful as we went up – huge trees, alpine flowers and small lakes, huge rocky crags with quite a lot of snow remaining, and wide views of the desert behind us. We took the camping stuff up to a most beautiful green plateau at about 10,500 ft and laid out our bedding on a patch of sand under an overhanging rock. Then, after making coffee and eating we went on to a plateau at about 11,800 ft below Mt. Whitney, on a rocky crag sticking up between several big lines of mountains, with a superb view of the whole formation. The weather was, of course, quite perfect and can be relied upon – a thing unknown in the Alps. We got down again to our car shortly before dark, made a huge log fire, and cooked an excellent supper over a minute petrol stove. The expert campers assured me that however many sleeping bags one takes one must expect to be cold at night. So we banked up the fire, arranged stone walls around it and had it in such a position that a large rock reflected heat back towards the rock we slept under. In spite of air temperatures below freezing, we were all four too hot! Perhaps the most memorable moment of the trip was opening my eyes at dawn to see the great rocky mountain straight in front of me, with its thin pink line of sunshine along its summit.
>
> We got down to the car about 10 a.m. and were washed, and dressed and talking viruses in the lab by 5 p.m.!

As Rosalind left Pasadena and headed for San Francisco, she wrote home from a bus stop, 'I have completely fallen for Southern California.'

* * *

That may have been her last untroubled day. At some point while in California, she was struck by sharp pains in her lower abdomen. The attack lasted two days. She dismissed it; she had always been healthy apart from a bout of jaundice when she was twenty-two. When the sharp pains came back – not once but twice – she went to an American doctor who gave her some painkillers and told her to see a doctor as soon as she got back home. It was exactly the advice she wanted. She was determined to wring the most out of her valuable trip, including return visits to Woods Hole and New York, and she did not want the inconvenience of going into an American hospital. Whatever was wrong with her could wait.

While in Berkeley, she celebrated her thirty-sixth birthday. She had decided before she left to buy a car, and the gift of money from her parents announced in birthday letters and a telegram on 26 July would go to buying things for the car 'or alternatively on bribing an examiner to let me pass my test'. (She had no licence yet. Her cousin Ursula, an experienced driver from military service, had accompanied her as she drove twice around Hyde Park Corner trying to exit.)

Rosalind's three weeks at the Virus Laboratory at the University of California at Berkeley were a let-down after the exhilarating time she had had so far:

> As on my last trip, this is far the least friendly of the places I visit. Everybody seems to wonder why I have come and to behave as though I hadn't . . . It is probably the biggest and most important virus research establishment in the world but both as a unit and as individuals they take themselves far too seriously. But I think I am managing to collect some useful information to work on when I get back. I'm staying in a somewhat sordid hotel because the person who booked it for me assured I'd be short of dollars, which I'm not, and it didn't seem worth the effort of moving.

Fraenkel-Conrat's wife, Bea Singer, a biochemist, noticed her air of sadness. Singer attributed it to a feeling that her work had not received due recognition. She invited Rosalind to their home and, like so many others, was impressed by Rosalind's rapport with their small children.

One reason for Rosalind's malaise was that Don Caspar would not be joining her as they had planned. His father had died suddenly of a heart attack and Caspar would be remaining with his family, who lived in Colorado Springs. Yet the Berkeley visit was not entirely devoted to laboratory work. 'The Cousinhood' extended to America's west coast and brought her one day to the home of Rosemary Montefiore: 'The house – quite fabulous – a concoction of Hollywood, 19th century English and palatial Mediterranean – and quite out of keeping with the way people live here but a most beautiful position in the Berkeley Hills overlooking San Francisco Bay.'

In Berkeley Rosalind also acquired two new adoring friends whom she met at a party. Ethel and Irwin Tessman were phage geneticists who had been graduate students with Don Caspar at Yale; they were at Berkeley learning new techniques in Gunther Stent's laboratory. They stood in for the absent Caspar and took Rosalind on long outings to see the giant redwoods and along the coast, where she came across a wild rocky beach

> like a much enlarged Cornwall, with the wild life much enlarged too – mussels up to 5 inches long, enormous bulbous starfish and seaweed like trees. The sea was not rough but there was obviously a strong outward drag which would make swimming dangerous. All this seemed like a natural consequence of the Pacific being so much bigger than the Atlantic. Big sea lions too on an island covered with cormorants, and a river estuary with dozens of herons.

Once more Rosalind inspired rapturous adjectives from American admirers. The Tessmans found her 'joyful, full of optimism, physically very beautiful, delightful and generous'. So smitten were they that Ethel Tessman determined to make a match between Rosalind and Caspar, sensing that the two were attracted to each other. Ethel took her clue from the way 'Rosalind used to glow when she spoke about Don'.

Rosalind did not conceal from Don Caspar her disappointment at their broken plans and tried to work out a way in which she could see him again before she returned. In her letter of condolence, she said so three times – a forceful and open expression of feeling for the bottled-up Rosalind.

> Berkeley
> July 30 [1956]
>
> Dear Don,
> I am so very sorry to hear your bad news. I've been hearing such a lot about your family from the Tessmans in the last week, I almost feel I know them – and so appreciate the more what a sad occasion this is for you.
>
> I am sorry I shall not now see you here. I shall probably finally leave New York around Aug 20, so perhaps I shall still have a chance of seeing you in the East? I shall probably leave here on Saturday and spend a few days sight-seeing on the way East, and then divide my remaining time between New York and Woods Hole and/or Boston (depending on Jim, Alex, etc). Anyway, after this week my address is c/o Bea. I hope I shall see you.

She went on to enthuse about 'some *very* impressive Steere electron micrographs of TYM showing beautifully regular knobs' and marvelled at how long the Virus Lab had had them 'before Ciba! [the Ciba foundation's symposium in London in March] – and Robley "wondered whether they were interesting"'. She thought she had convinced him that they were.

A few days later she was able to write to Aaron Klug, 'I'm going to see Don after all.' She also told him that she had postponed her departure by a week to 21 August, that Fraenkel-Conrat and Robley Williams had found that the reconstituted virus was infective and that, looking forward to getting back, she hoped that Maurice Wilkins would not ask for the return of the King's camera they had been using.

Matchmaking Ethel Tessman did her best to get Rosalind to Colorado. She put Rosalind in touch with a young scientist, Seymour Lederberg, brother of the molecular geneticist Joshua Lederberg, who was driving cross-country with a friend and could give her a lift as far as Denver. Before leaving Berkeley, Rosalind was cheered by some last-minute results which made the three weeks of hard work in the important research centre worthwhile. The drive, which she had never done, through the moonscape and cathedral rocks of Utah and Nevada, gave her an opportunity to resume traveller's tales for her parents, with just a veiled reference to the 'friend from the East' whose father had died and whom she was travelling a thousand miles to see. (Good at keeping things from her parents, she never mentioned her episodes of severe abdominal pain.)

> Then, in spite of everything, the family urged me to come here for a few days on my way back. Finally I was driven, in 2 days, from Berkeley to Salt Lake City by two people who were driving from coast to coast, spent Tuesday night in SLC and flew on here (in Denver) Wednesday morning. Monday's drive started hot and dusty and then got very beautiful as we crossed the High Sierra mountains and picnicked by Lake Tahoe – the biggest lake after the Great Lakes, over 6,000 feet up, at sunset. We then drove on until midnight and slept out in sleeping bags in the desert, somewhat uncomfortably, by the road-side. The next day's drive was somewhat mountainous but interesting, across desert mountains of Nevada and Utah, and the extraordinary Great Salt Lake desert, which extends for about 100 miles as a great flat white sheet of salt resulting from the evaporation of salty water which seeps up from below. At the junction of these two kinds of desert is a little town consisting of a big line of 'gas stations', a railway yard, some military officers' houses and called surprisingly Wendover! At one stage of the journey we drove nearly 60 miles looking for a tree, stone or other object which would provide shade for 3 of us to sit in and eat lunch – and finally found a

shack put up for the purpose. There was no natural shade.

Here I'm being taken around the country on leisurely car rides which are a pleasant change from the rushed driving of Monday and Tuesday and entertained to truly magnificent American-style picnics, the first of them on a hill just behind the Rocky Mountains from which we watched the sun set behind the Rocky Mts while we ate.

Sunset over the Rockies: that Rosalind should have known Caspar well enough to have visited him in his home at a time of family bereavement says something about their closeness, even if they did talk about electron micrographs of TMV. The relationship was chaste. Caspar was in awe of her as an older woman of formidable ability. Yet, what not long after she told Dr Mair Livingstone (the good friend to whom, with a new baby, she had loaned her flat) suggests that the Caspar friendship seems to have been the nearest, Mering apart, Rosalind came to being in love. Anne Sayre's 1974 biography of Rosalind ventures that during her 1956 visit to the United States she met a man 'she might have loved, might have married'. Sayre went further in a private letter, claiming that Rosalind was bespoken in the summer of 1956 to 'a delightful man who was also an ideal match'. For Irwin Tessman, who had known Caspar since Yale, 'Most telling for me was the enthusiastic way he talked about her' and added that he found it unlikely that Caspar could have had a close working relationship with such an attractive woman and yet not have been in love with her.

Ethel Tessman kept what she considered a burning secret (out of respect for Caspar's later happy marriage) for twenty years until Dr June Goodfield, researching a proposed film about Rosalind, raised the matter, to be told, 'I've been waiting twenty years for someone to ask me that question!' Suffice it to say that three women close to Rosalind believed that she fell in love with Don Caspar that summer.

* * *

Rosalind concluded her two months in America with a lazy weekend in 'perfect weather' at Woods Hole. Caroline and Francis ('Spike') Carlson took her for a boat ride past Penzance Point so that she could see the tiny peninsula's great shingled mansions from the water. The Carlsons found her charming, fun and '*very* good looking'. Leaving Woods Hole for New York for talks with the Rockefeller Foundation, and more lab visits, she alerted her mother to a new delay of her return:

> I've made another and final postponement of my departure. I have reset Thursday Aug. 23, arriving in London (at the airport) at 9.20 a.m. on Friday. Unless we arrive very late I shall go to the lab on Friday – this is why I chose to arrive Friday rather than at the weekend.

The postponement may have been unwise. At the beginning of the last week of her trip, while in New York, she had trouble zipping her skirt. She had always been trim, comfortable in trousers, shorts or slim skirts, yet suddenly her midriff began to bulge. Nonetheless, she kept to her timetable and on Thursday 23 August boarded the plane at Idlewild Airport. The Tessmans who were attending a conference not far away at Cold Spring Harbor Laboratory, came to see her off. She arrived in London next morning, and went straight to the lab as planned.

But she knew she needed medical attention. She went to see Mair Livingstone, who was in general practice in Hampstead, and listed her symptoms. 'You're not pregnant?' asked Livingstone. 'I wish I were,' was the answer. Livingstone instantly suspected ovarian trouble. However, she simply told Rosalind that it sounded like a cyst and she ought immediately to see the University College Hospital doctor (Dr Linken) who had examined her four months earlier.

Linken also thought of the obvious first. But after vaginal and rectal examinations, he wrote in his note to the surgeon who would see Rosalind the following day, 30 August, 'there is no reason to believe there is a pregnancy – ectopic or otherwise'.

Professor Nixon, who saw her, felt the large lump in the right pelvis, told her she had to come into hospital, and marked her case file in large letters, 'URGENT'.

Clearing her desk prior to admission, Rosalind wrote several long letters. To Wendell Stanley at Berkeley she summed up her three weeks at the Virus Laboratory at Berkeley. At first she had prepared orientated gel specimens for X-ray from reconstituted TMV and from TMV protein combined with Ochoa polymers (a synthetic pseudo-RNA, in which all the bases were the same), but the experiment failed. Eventually, she found that treatment with ribonuclease allowed her to obtain the normal optical and flow properties of the virus. This led her to conclude that the original specimen had been gummed up with excess RNA in preparation. Now she could see a way to get it ready without distortion for X-ray analysis.

To Dr Pomerat at the Rockefeller Foundation, she wrote, with gratitude, that the pace of American virus research was so rapid that 'a large proportion of the most interesting results are still unpublished, and it is only by personal contacts that one can even form a general impression of what is going on'.

Three days later Rosalind was admitted to University College Hospital as a National Health patient for investigation of 'an abdominal mass'. The operation on 4 September revealed, Professor Nixon reported to Dr Linken, 'the findings are most unfortunate': not one tumour but two. In the hearty terminology favoured by Harley Street medical men of the day, Nixon described the large lump in the position of the right ovary as the 'size of a croquet ball', and the cyst found on the left ovary 'in size equal to a tennis ball'.

The surgeon was less metaphorical when he rang up Rosalind's parents. Muriel Franklin answered and was told, 'Your daughter has cancer.'

EIGHTEEN

Private Health, Public Health

(September 1956–May 1957).

BEFORE HER OPERATION, Rosalind gave consent for a hysterectomy if one were necessary. She knew that she might be signing away her chance to have children and noted wryly that she had been put in the Obstetrical Hospital. Perhaps when her brother David came to collect her, he might think she had had a miscarriage or an abortion. When she emerged from the anaesthetic, however, she was relieved to hear that her womb was still there and a part of one ovary. But not for long.

Almost exactly a month later she was back in University College Hospital (this time as a private patient) for a second operation – ostensibly for an infection, actually for a hysterectomy and removal of remnants of the left ovary. This was done and no further cancer was found.

Anne Sayre, who was staying in Rosalind's flat, answered the telephone one day, to hear Jacques Mering calling from Paris. He was startled that Rosalind was not there, and upset to learn that she had undergone further surgery. As Anne struggled to explain in French, Mering became very agitated. Later, when Rosalind learned of the call, she too became very emotional, then confessed she had been very much in love with Mering.* She gave no details but Anne gathered that it was a matter of deep feeling on Rosalind's part and a source of unhappiness, because – so Anne believed – Rosalind was incapable of any prolonged

* The Franklin family disagree once more with Anne Sayre's account.

happy extra-marital relationship. Yet Anne, who knew Mering from Paris, recognised that he combined intellectual strength with 'a truly immense personal attractiveness and a European attitude and psychology'. Anne was aware too that, since Rosalind had left Paris, she and Mering had seen each other several times a year, and corresponded. (Mering later destroyed Rosalind's letters.)

If Rosalind knew that ovarian cancer is called 'the silent killer' because it is often well advanced by the time it is diagnosed, she put the thought aside. She was optimistic. Cancer had been found, she understood, but in a contained form that surgery had removed. Convalescing at her parents' home in north London, she loaned 22 Donovan Court to another friend, Anne Sayre having returned to America. 'Use my flat,' Rosalind said to Jean Kerlogue, saying she was recovering from 'just a little operation'. But Alice Franklin, Rosalind's aunt and neighbour from whom Jean got the key, indicated that the condition was more serious than Rosalind let on.

Rosalind was still staying with her parents when she wrote to Anne to tell her to stop worrying:

> everything is going very well, and I expect to be fully back to normal some time next month. I've been out of hospital a week, and from here I manage to keep in touch with what is going on in the lab and even to catch up a bit on writing things up. My mother really seems to have understood that I not only don't mind being left alone at times but positively like it, so things are a lot easier than last time. Next week I'm going again to Cambridge to stay with the Cricks – it's really rather hard on relatives and friends who offer to take charge of my convalescence to have to put up with me twice in a month, but I've promised them I won't do it a third time. After Cambridge I shall probably stay a bit with a brother or two before going back to Donovan Court.

At home, she was silent for long periods. Her parents were not surprised; they were accustomed to her uncommunicativeness. She would sit with her knitting or sewing, then suddenly say, 'I think it is my bedtime,' and take herself away at an early hour. Ellis and Muriel sensed that, in fact, she was getting very little sleep. As soon as she could, she moved out, saying she was sorry she had not been better company.

The emotional distance between mother and daughter, so long filled by Nannie, had left Muriel and Rosalind without that shared understanding many mothers and daughters carry throughout their lives. Muriel used to weep and get terribly upset if Rosalind did not eat her lunch or take a rest. Rosalind's reaction to undisguised maternal anxiety was to retreat to others who did not try to stop her doing as she wished.

The Cricks certainly did not fuss. Francis Crick never asked the nature of her trouble; he told a woman scientist who inquired that he thought it was something 'female'. The scientist was Dorothea Raacke, an American marine biologist and friend of Crick's. Raacke first met Rosalind in Cambridge, at Crick's table in The Eagle, and asked her how she liked to be known. 'I'm afraid it will have to be Rosalind,' was the reply, pronounced in two quick syllables; then, with eyes flashing, 'Most definitely *not* Rosy.'

'There was, thought Raacke, with the Shakespearean and operatic Rosalinds in mind, 'nothing at all romantic about her despite her beauty'. Raacke observed that Rosalind was more timorous and less assured than her assertive reputation suggested, and that she relied heavily on Crick for advice and encouragement. Raacke sympathised. From her own experience, she saw that women had a difficult time in academic life, particularly in science, and accepted it as given that in science a woman was judged and criticised much more harshly than a man and got less acknowledgement for work well done.

The two women became good friends and Raacke stayed many weekends at Rosalind's London flat. The topic of men was one Raacke felt she could not broach. Herself engaged to be married, she was longing to ask Rosalind if she had ever faced the conflict between marriage and her ambitions. But the wall of reserve was too strong. Nor did they discuss her illness. It was nearly a year before Rosalind confided that she had had a hysterectomy.

Rosalind was staying with the Cricks during the Suez Crisis of October–November 1956. Britain, France and Israel were embarked on the futile attempt to seize back the Canal Zone nationalised in the summer by the Egyptian president, Gamal Abdel Nasser in protest against the decision of the United States and Britain to abandon their promise to finance Egypt's construction of the Aswan High Dam. Thus preoccupied, the major Western allies did little to help the Hungarian rebels who had risen up against Communist rule. Stalinism had begun to crumble eight months earlier when Khrushchev denounced Stalin at the twentieth congress of the Communist Party; his speech set off a wave of unrest in the satellite countries of Eastern Europe. When this dissent led to an armed uprising in Hungary, Soviet troops and tanks moved in and took control – an invasion that might not have been attempted but for the Suez diversion of Western attention.

Returning to Birkbeck, Rosalind watched with wry amusement the consternation of the college's many Communists at the Soviet Union's brutal action. 'Reactions in Birkbeck to Hungarian things were quite interesting,' she wrote Anne Sayre; 'nearly all the Party members were quite shaken, though I don't know any who left the Party.' Bernal, she observed, was fairly rattled, and on the rare occasions she heard him refer to the event, he got 'emotional and confused', but soon 'was again talking in the old Party clichés'.

As she had planned, she bought a car – a two-door grey Austin. For her colleagues, this was another quiet sign of the income gap between her and the rest. If they owned a car at all, and very few did, they had old models whose parts they were constantly tinkering with in the Birkbeck workshop. Rosalind's car was new.

In 1956 cancer was not a word freely uttered. Obituaries

spoke delicately of people dying 'after a long illness'. Rosalind's team did not know what had been wrong with her, simply that she seemed restored to health. From Harvard, Jim Watson wrote Aaron Klug, 'I have heard rumours that Rosalind was not well. I hope she recovers soon,' and asked him to give her his wishes for a speedy return to health.

Her colleagues were glad to have her back. They were working on the comparison of plant viruses, with the amicable cooperation of Maurice Wilkins. Wilkins penned a note to 'those in charge of cold rooms & fridges' at King's College: 'Please give Dr Klug every assistance so that he may search for a bottle of TYMV [turnip yellow mosaic virus] which we wish him to have.'

With her formidable control over intrusive emotion, Rosalind's main concern was the future financing of her research group. She had developed a well-knit and loyal team but time was running out on their support. Sir William Slater had let it be known that funds from the Agricultural Research Council were unlikely to continue beyond the end of April 1957. There was no more in sight from any other British source. A month back at work, Rosalind submitted a clear, well-thought-out application to the Public Health Service of the US National Institutes of Health. Formally explaining her reasons for seeking help from abroad, she spelled out that her British grant had been decreased, that the collaboration between herself and Dr Aaron Klug was proving fruitful, and that the well-balanced research team they had built up should be allowed to continue for at least five years. Their project, moreover, was unusually international in character.

Under 'Significance of this research', she summarised three years' investigation into the combination of ribonucleic acid (RNA) and protein:

> Recent work has shown that there is a close similarity between the small RNA-containing viruses such as poliomyelitis and the 'spherical' plant viruses with which our work is largely concerned. Moreover preliminary work on

> cytoplasmic nucleoprotein particles from various sources indicates some degree of structural resemblance between these and the small viruses.

The information into the internal structure of viruses gained by X-ray diffraction studies, she emphasised, could not be obtained by any other available technique. Accordingly, she requested $28,345 for the year beginning 1 October 1957 – roughly £10,000.

Illness pushed to the back of her mind, anxious about the unsettled future of her research group, she ended a difficult year with a trip to Paris, then to Strasbourg to see the Luzzatis, now the parents of two small children. Rosalind was amused to watch the children watching her: 'The thing that impressed Anne (the eldest) most about me was that I ate an egg for breakfast – she was convinced that I'd made a mistake about the time of day.' During the visit, which included a walk in the Black Forest, Rosalind never spoke of the nature of her illness. She was back in the lab by 9 January.

Ever her booster, Bernal wrote to the ARC's chairman, Lord Rothschild, pointing out the great progress Rosalind's group had made:

> The results which are beginning to show the precise relation of nucleic acid to the protein component are just now at the very centre of interest of biological structure analysis and are already beginning to tie up with the structure of such components as microsomes and chromosomes.

Rothschild, replied to 'Dear Sage', signing himself 'Victor', that arrangements would be made for Miss Franklin which he hoped would be satisfactory.

For the few months of 1957, Rosalind regretted that she had energy for working half-days only. Even so, she went into to the lab every day, preparing her specimens, drying her gels, turning them into crystals for the X-ray camera, comparing the evidence from different varieties. The papers poured out. She had published seven in 1956; six were on their way in 1957, mainly about viruses and co-authored with her colleagues, Aaron Klug, Ken Holmes and John Finch. An exception, which appeared in *Nature*, concerned oxidation in carbon and was written with her doctoral student, James Watt. An important long article on the turnip yellow mosaic virus, signed by Franklin, Klug and Finch, had been intended for *Nature*, but, at Crick's suggestion, sent instead to *Biochimica et Biophysica Acta*. 'I feel in the long run,' Crick said, 'it does people more harm than good to publish too much in *Nature*.' (A condensed version of the paper was sent to *Nature* anyway.)

At intervals, Rosalind went back to University College Hospital for a checkup. On 20 February, a follow-up examination reported that she was feeling well generally, although hot flushes through the night prevented her sleeping. Confidently Klug wrote to a biochemist in Kansas, 'Rosalind Franklin is well again and back in the laboratory.' From what Klug could see, 'She looked okay.'

Nothing had been heard from the US Public Health Service when, in mid-April, the Agricultural Research Council (true to Rothschild's word) renewed its grant – but for only one more year. This was to be the *final* – underlined *final* – year: £4,300, to end in March 1958. After that, the ARC hoped that American sources would provide support. In a frosty coda, the council's governing committee observed that it was 'inappropriate' for a worker of Miss Franklin's seniority to be carried on an annual grant. They were concerned at the absence of any steps, apart from the application for American funds, to establish the research on a more permanent basis. Miss Franklin, they felt, should either seek a post in a university or a research institute. They understood that she was not willing to do so.

Or able. The illusion of normalcy was shattered at the end of April when Rosalind was readmitted for two days to University College Hospital, very distressed, with profuse bleeding from the rectum. Two weeks later she was back in hospital, complaining of abdominal pain. An examination revealed a new mass on the left side of the pelvis. With the endless rounds of internal explorations, investigative enemas, palpating, probing and discussing, a very private woman was private no longer.

Rosalind asked her surgeon for an honest prognosis, and he gave it – frank and discouraging. He urged her to seek the comforts of religion and be grateful that she had time to prepare her soul. She was furious – with him, for despair; with science, for not yet having discovered a way to halt the progress of her disease until a cure could be found. She chose to remain hopeful; her new treatment was to be cobalt radiotherapy. Starting in mid-May, she disappeared from the lab from time to time and returned. True to form, she never told anyone where she was going.

Concerned with the personal futures of her young team as well as for the future of her research, she put enormous motherly effort into investigating sources of finance for them all. Cheering news came from the secretary of the Medical Research Council who thanked Sir Lawrence Bragg of the Royal Institution for letting him know how highly he thought of Miss Franklin's work. The MRC would consider supporting two of her students when the ARC money ran out.

New research material arrived for crystallographic analysis. The Sloan-Kettering Institute for Cancer Research in New York sent her some rat liver nucleoprotein, frozen and stored in a thermos bottle, and carried across the Atlantic by Francis Crick. Rosalind, standing in a 'cold room', packed the substance into capillary tubes for X-ray photographing later. At the same time, she was under pressure to build models of her tobacco mosaic and turnip yellow mosaic viruses; those planned for the Brussels World's Fair would be five feet high. Professor J.W. Moulder of the University of Chicago's microbiology department, who was in charge of the international science section at the fair, enthusiastically urged her to make them as large as possible for maximum visual effect. The models, he said, were to be a highlight of the general virus exhibit and had to be ready, with explanatory material understandable by the general public, by November.

With such a full desk and diary, Rosalind was too busy to die.

NINETEEN

Clarity and Perfection

(May 1957–April 1958)

NO ONE WHO SAW ROSALIND in public in the summer season of 1957 would have guessed she was living under threat. Hiding any pain or weariness, she stood on her feet for long hours in evening dress, demonstrating and explaining her work at exhibitions. One week into cobalt therapy, she presented her models at the Royal Society's *conversazione* as part of a display on the application of X-ray analysis and electron microscopy to rod and spherical plant viruses. Her parents came and found her looking lovely, gay and happy. Three weeks later, on the evening of the day of her last cobalt treatment, she put on a red silk Chinese blouse for a Friday evening discourse at the Royal Institution – a pre-Victorian ritual at which an invited black-tie audience listens to a lecture, then strolls through displays in the library. Accompanying the talk – 'Observations on the Architecture of Viruses' – by Robley Williams of the Virus Laboratory at Berkeley, were four exhibits: two by himself and Kenneth Smith, one by Crick and one by Franklin and Klug.

Francis Crick continued to be a friend and uncompromising critic. 'Rash in the extreme' was his judgement on a Franklin–Holmes paper on the use of heavy atom substitution to determine the internal measurements of the tobacco virus. Listing his objections, Crick advised more research and warned of the dangers of publishing wrong information: 'There is absolutely no urgency to publish a paper at the moment. Nothing is lost by postponing publication, whereas if you publish it and it turns out to be wrong your whole research programme will be discredited.'

No urgency, of course, unless you were dying. But that is not how Rosalind saw it. She wanted to submit the paper she and Ken Holmes had written, to *Acta Cryst*. She took Crick's comments to heart, as she undoubtedly did also those appended by Max Perutz whose wise observation applied to more than crystallography:

> One so easily persuades oneself of the rightness of a dubious solution and then spends one's time in fruitless search of supporting evidence. As Francis has said, this is an occupational disease among protein crystallographers. It can be kept at bay only by frequent cold showers.

Don Caspar reappeared in her life. He was in Cambridge, with his mother, for the summer of 1957. Rosalind had got along extremely well with the widowed Mrs Caspar when they met the previous August in Colorado and their friendship resumed when Rosalind came to visit Francis and Odile Crick. In subsequent weeks, Rosalind had Mrs Caspar to stay at her London flat; one day, for mother and son, Rosalind organised one of her favourite treats – a picnic in Richmond Park.

As Rosalind's health seemed to improve, so did the outlook for her virus group. On 9 July she heard from an official at the US Public Health Service in Bethesda, Maryland, giving her informal notice that her research grant had been approved by the National Advisory Allergy and Infectious Diseases Council. J. Palmer Saunders sent the news 'preliminarily,' he explained, 'because of the unfortunate delay you have encountered previously'.

However, the news was not greeted with universal rejoicing at Birkbeck. There was a feeling that the American grant, £10,000 a year for each of three years, was too much: £10,000 was enough to be spread over three years rather than be paid as an annual sum.

* * *

Two conferences and a holiday in Europe in mid-summer: why not? Rosalind was feeling improved even though re-examination in July showed that the pelvic mass on the left side was still there and her parents asked for a second opinion. When the hospital advised that Rosalind remain near London just in case some intestinal obstruction should occur, she took the advice instead of her friend Dr Mair Livingstone, that there were doctors on the Continent should she need one. However, Rosalind was not oblivious to her condition; she told Livingstone that in the United States the previous summer she had met a man she might have loved, even married, but whom she had put out of her mind because of her illness.

She and Caspar attended a conference on polio in Geneva, where the star was Dr Jonas Salk, discoverer of the polio vaccine in 1953. Afterwards, with Caspar, his mother and Richard Franklin (no relation – rather a friend of Caspar's from Yale who owned a Volkswagen) she went to Zermatt. Mrs Caspar photographed the trio silhouetted against the great peak of the Matterhorn. Rosalind did not mention her illness, and seemed well, even if she did not have the strength she had had the previous summer in the Rockies. For her, the Alpine excursion in the company of a close friend was the highlight of her summer: 'a glorious weekend in Zermatt', as she summed it up for Anne Sayre.

Leaving Geneva, she attended a protein conference in Paris, then embarked on a long driving tour through northern Italy, with her sister Jenifer and a friend. She would have preferred something more rugged:

> This was my first continental holiday by car, and confirmed my suspicion that cars are undesirable on holiday. Italy was wonderful, as always – the Dolomites, Verona, Ravenna, Pistoia, Lucca, a delightful day with the Luzzati family near Viareggio and the sea at Portovenere. But travelling around in a little tin box isolates one from the people and the atmosphere of the place in a way that I have never experienced

> before. I found myself eyeing with envy all rucksacks and tents.

She was in no way an invalid on the trip, made in Jenifer's Morris Minor which had been flown across the Channel in the primitive air-lift operating from Lydd on the Kent coast. And her good humour made a strong impression on Luzzati's mother, who was convinced that any woman scientist would look and behave like a suffragette. She was surprised and pleased, he recalled, to find out how attractive and easy-going Rosalind was.

Polio was her new interest. In mid-1957 she began to study the virus – rather, to try to. It was spherical, morphologically close to the turnip yellow mosaic virus. When two scientists at Berkeley offered to supply her with polio crystals, she wrote to ask Bernal's permission. He, in turn, sought the Master's: 'In view of the extremely small amounts of infective material with which she will be dealing, and the very careful precautions she will be taking, there could be no objection to the research being carried out.'

Bernal's plea was thwarted, not by his usual adversary, the Master, Dr J.F. Lockwood, but by the staff. Some felt that Birkbeck was so old and filthy that if it were cleaned up it would fall down – hardly premises in which to store a dangerous material. The physicist Werner Ehrenberg was particularly alarmed, as he himself had been crippled by polio. At one point, Rosalind stored the crystals in her parents' refrigerator. As she put them in a thermos flask, she said to her mother, 'You'll never guess what's in there. Live polio virus.'

In time, the virus was housed at the nearby London School of Hygiene and Tropical Medicine. From the School, Professor E.T.C. Spooner commiserated with Bernal in having failed to dispel 'the fears of some of your people; the magic word "poliomyelitis" has such a terrific emotional build-up at present that one cannot expect normal critical faculties to work':

> I am myself quite sure, as I think you are, that the risk associated with Dr Franklin's proposed work is quite negligible, smaller probably than that which anyone takes who subjects himself to any kind of injection, for instance. Of course there is a risk; as there is that a house will burn down, or that a meal in a restaurant will give you botulism; or there is the far greater risk of walking across a road. But that is not the point; and I doubt if anything I can say will alter the position. However, I do hope it will be possible for Dr Franklin to do this work. To have to confess to our American friends (who already laugh at our timidity over the polio vaccine) that it cannot be done because of the risk that the tube might break would, in my view, be extremely humiliating.
>
> We in my department will gladly do anything we can to help.

With the virus safely housed, Rosalind took it for X-raying to the Royal Institution. Klug and Finch, who were working with her, chose to take the vaccine, but Rosalind decided against it. Her decision has been interpreted as fatalistic courage; as she already had a terrible disease, she could risk another. It seems more likely, however, that Rosalind, rather than manifesting an indifference to survival, was as conscious as was Professor Spooner of the low probability of an accident. She knew that she was not going to drop or break the test tubes. Proudly she added a new title to her curriculum vitae: 'From Oct 57 Project Director of a US Public Health Service grant for research in molecular structure of viruses by X-ray diffraction.'

By November, the lull was over. Her stomach had swelled up again, she told Anne Piper over the telephone, and she needed to have it drained. It seems to have been at that point that she became a patient at the Royal Marsden Hospital, just around the corner from her flat in Drayton Gardens. Her admission to the Marsden, an institution specialising in cancer treatment, was the first evidence her staff had of what her illness really was. At the Marsden, she was one of the first to receive chemotherapy. Alice Franklin, her aunt who lived very near, marvelled at her courage, also her faith, each time, that the latest treatment would hold the disease in check until a cure could be found. In a rare willingness to talk about her work with her family, she offered to show her Aunt Alice some of her drawings and diagrams.

Rosalind liked her doctors at the Marsden, especially her case supervisor, the well-known Dr David Galton of the Chester Beatty Institute, part of the Institute of Cancer Research, who would sit and talk with her and in whom she had utmost confidence. With her radiotherapist too, she had good rapport, asking questions and getting intelligent answers.

Visitors to her room in Granard House, the Marsden's private wing, found her in a wine-coloured dressing gown, taking radioactive liquid gold in a tilting bed, with a television set in her room, and Jenifer or one of her brothers usually at her bedside. Nannie came to see her. After a week, however, she was still in pain.

Into her room one day walked Peggy Clark, who had read science with her at Newnham. Peggy was now, under her married name of Dyche, a medical physicist on the Marsden staff. She had seen a parcel addressed to Dr Rosalind Franklin and wondered if it was her old college friend. Not having seen Rosalind since Paris, Peggy found her looking reasonably fit and cheerful. Rosalind was happy to talk; she complained about the incompetence of the doctors at University College Hospital and spoke of her hope that the chemotherapy would be successful. She described with pride her virus model to be exhibited at the Brussels World's Fair. In subsequent days, when she got out of bed, Rosalind would occasionally wander into the hospital laboratory to chat with Peggy and other staff about the radioisotopes they were dispensing.

In and out of hospital at the end of 1957 and the beginning of 1958, when not well enough to live by herself in her flat, Rosalind stayed with her brother Roland and his wife Nina at their home on Aylmer Road in East Finchley. She listed Roland as 'next of kin' on her hospital admission form. Her mother would ask, 'Why are you going to Nina's?' but the rest of her family understood. Rosalind positively enjoyed staying with Roland and his wife, who had several small and lively children. Her room was ready and her bed warmed for her whenever she wanted to come; she could rest when she wished, and otherwise play with her nieces and nephews. In Roland's view, his mother was over-solicitous and his father seemed unable to cope: 'I think she just felt more comfortable in our house. We never discussed her illness. It's not a confiding sort of family; we were very private and we would not share emotions or thoughts or anything with each other at all.'

Perhaps not, but her brothers, as so often before, were emotional props. She did have her dark moods: the thought of dying with her work unfinished made her furious and depressed; she would ring Colin or Roland in the middle of the night if she needed to talk. All three brothers were named as executors when, on 2 December 1957, she made her will. None of her relations were named as beneficiaries of her modest estate – not surprising, in an affluent family whose financial affairs were well-organised. Instead, she chose Aaron Klug as her principal beneficiary, to receive £3,000, plus her Austin car. Next on her list were two close women friends who had children to support: for Dr Mair Livingstone, £2,000; for Anne Piper, £1,000. Last, 'my old Nurse Miss Griffiths' was to have £250. The residue of the estate was to be distributed by the executors to charities 'of which they think I would have approved'.

The sums involved were not small at the time to people without savings or capital. The legacy to Aaron Klug, as seen by his friend, Dan Jacobson, was a brilliant and sensitive stroke. Jacobson imagined Rosalind saying to herself, 'What does Aaron need?' She knew that the Klugs, with a small son, were very short of money and that he was considering going back to South Africa. Her gift was thus well in the Franklin family tradition of enlightened philanthropy.

*   *   *

Still in hospital over Christmas, she laughed when some worthy ladies made knitted presents for the patients, and hers turned out to be a pullover with no hole for the head. She was not *that* ill, she joked. By early January 1958 she felt well enough to go on an outing with her cousin Ursula. Rosalind said she wanted some French food so Ursula drove her to a good restaurant in Richmond; she also enjoyed a walk in the park. Later she described the day out in a letter in French to Adrienne Weill, with the vividness ('*La journée de hier a été magnifique*') she had put in her holiday letters about her Alpine adventures. Indeed, she felt so well after her excursion, she said, that it seemed ridiculous to return to the hospital to play the invalid. She decided to negotiate her release, continue the treatment as an outpatient and begin making appearances at the lab. From then on, she said, she would, little by little, return to normal life.

A few days later she did as she wished. Leaving hospital in mid-January, she signed a three-year lease on her flat and wrote in French to Luzzati (still scrupulously using the '*vous*' form of address) that she had been living in idleness at her brother's (she described herself as having been a '*parasseuse*' – lazybones) but now that she was back at the lab she was busy with administration and editing, but looked forward to starting new work once she got the right crystals. On 25 February she was notified of a pay rise (thanks to the promised American money) and a college appointment. Her rank was to be 'Research Associate in Biophysics', and her salary would rise to £1,700 a year, plus £80 London allowance, to take effect from October. It was good news, and would have been better were not she and Klug still waiting for the previous salary rises promised, but never paid, the previous October. Still fighting on his behalf, she added her name to his in a joint letter asking for a rise in his salary in line with the general London University increases.

* * *

Welcome as her overdue recognition was, it was hardly the kind of prominence to put a scientist in line for election to the Royal Society. Francis Crick and Maurice Wilkins, on the other hand, had advanced considerably in reputation since the double helix days of 1953. From the Royal Institution that spring, Sir Lawrence Bragg, in a letter headed 'Confidential', wrote to Max Perutz at the Cavendish Laboratory at Cambridge to discuss candidates for fellowship in the Society. Crick, they agreed, took precedence over Wilkins. Bragg said, 'If you feel you want decisively to run Crick as your horse, perhaps Himsworth [Sir Harold Himsworth, Secretary of the Medical Research Council] and I could put Kendrew up – I'll test his reactions.'

> I shall of course be glad to sign Crick's nomination. I imagine Rothschild will be very pleased to see it go forward. I agree with him that Crick ought to get in before Wilkins, I think R. was very worried because W. was so strongly pressed. I should like to see Kendrew go forward at the same time. I am going to have a talk with Himsworth shortly and will write again when I have done so.

Thus the network operated – gentlemen sorting things out among themselves. Rosalind's name never came into it. Crick and Wilkins became Fellows of the Royal Society in March 1959, Kendrew the following year.

Her work was interrupted in the spring not only by intermittent stays at the Royal Marsden but by practical frustrations with her working materials. The polio research had come to a halt because of an incompatibility between the virus and the capillaries in which the crystals were stored. She and Klug were waiting for some neutral glass out of which to make tubes that would not destroy the crystals. The tobacco virus work, on the other hand, was held up for the lack of a heavy atom derivative. Nonetheless, she and Klug submitted a detailed overview of disordered crystals in long-chain molecules (such as TMV and DNA) to *Transactions of the Faraday Society*.

In her spare time, Rosalind re-covered the cushions in her flat in bright colours and thought ahead to her visit to Brussels and to her summer travel. Writing to Don Caspar at Yale in mid-March, apologising for the fact that 'most of the winter has been more or less non-existent for me', she spoke of her gratitude to Klug for managing to do a good part of her work as well as his own, and she described their troubles with the polio crystals. With her holidays in mind, she suggested that Caspar join her for another excursion:

> I still hope to come over this summer . . . Although I am fairly well at the moment I have, in fact, been in hospital and out again since I got your letter. So it all depends how things go in the next few months.
>
> If I do come, I should like to come over for the month of August – Phytopathology is the last week of August, and I hope to go to Vienna in the first week of September. If you felt like a trip West during August it would be very nice if we could go over together.
>
> Best wishes,
> Yours,
> Rosalind

In her financial negotiations, she was increasingly successful. She won permission for her team to spend nine pounds each to attend the Faraday Society Conference in Leeds in mid-April. The expense was justified, she explained to Bernal, as all four were authors of papers to be presented at the conference. She also arranged that she and Klug could dip into the American funds to cover the costs of the trip to Vienna in August. (Although the US Public Health Service had made the grant, the money was issued by the College Research Grants Committee in its usual penny-pinching manner.) Those were small victories compared to the splendid news that ended her long anxiety about the possible dispersal of her group. Max Perutz came personally to Birkbeck and invited her and Klug to move their work to Cambridge, to the new laboratory being built to house his expanded molecular biology unit from the Cavendish. This transfer would take place about the time that their American grant ran out. Her team's future was secure at last.

Towards the end of March, Rosalind was putting in a full working day. Yet she was in a weakened state. She would crawl up the narrow stairs from where the X-ray apparatus was, to her office on the top floor, refusing all offers to be carried as her devoted young associates longed to do. Her courage drove Ken Holmes almost to tears. But she was determined to get the data for a paper on viruses, to be given with Klug and Caspar, at the Bloomington, Indiana conference in August. She had begun some new work on the potato virus and mounted some very tiny crystals and had managed to get a picture. Long columns of figures in her carefully dated notebooks reveal how much she packed into a day. She produced five pages of calculations on 28 March.

That week she went to dinner with Ken and Mary Holmes and they had a lively evening, Rosalind at her best, telling them stories about Paris and enjoying coffee ground in the grinder she had given them as a wedding present. Friday, 28 March, was Ellis Franklin's sixty-fourth birthday. At the family dinner celebrating the occasion, her mother, watching carefully, thought she ate well but looked ill. On Sunday Muriel was in the garden when Ellis came out and said to her, 'Rosalind's in trouble again.' His sister Alice had taken Rosalind back to the Marsden. When her parents went to see her, she asked not to be visited the next day – perhaps to spare herself the sight of her mother weeping.

Rosalind still expected to recover. On the table beside her hospital bed stood an invitation to a six-months fellowship in Caracas. Peggy Dyche, calling into the room regularly, knew, however, that her condition was hopeless. An operation at the Marsden showed that the cancer's extensive spread meant that nothing could be done. Nonetheless, Peggy kept up the pretence that Rosalind would get better 'when both she and I knew the worst.

I never forgave myself for not breaking through the facade.'

'Breaking through' would have taken some boldness, for she knew Rosalind very well: 'There is no doubt that she could be a difficult character – impatient, bossy, intransigent. She always went straight to the point and was seldom diplomatic. However, this was all because she had such high standards and expected everyone else to be able to reach her ideal requirements.' Peggy sympathised; she saw the same traits in herself.

Another visitor to the bedside was Jacques Mering. He was shocked at Rosalind's appearance and had trouble keeping his composure. Her hair was lustreless and she had wasted away to a skeleton. Yet she was cheerful and kept telling him ('with defiance', in his recollection) that she would get well. He thought to himself that she was defying death itself. When he left, he stood on a street corner and wept.

Mering's tears sprang not only from grief but remorse. He suspected (as he later told Anne Sayre) that Rosalind's attachment to him might have made it impossible for her to form other relationships, even to entertain the attentions of those other men he felt were paying court to her. Mering was conscious too that, as the older and more experienced of the two, he should have managed her feelings for him better.

One day when Peggy looked in, Rosalind told her that she could not move one arm; it felt paralysed. She was sure she had polio. Outside the room, Peggy commented to colleagues, 'There she is dying of cancer and now she thinks she's got polio.' But none of them realised that Rosalind had been working with the polio virus. Another day, when Rosalind seemed remarkably improved, with glowing eyes and smiling face, Peggy inquired and was told she had been put on heroin.

There was no cure. A new treatment – perhaps a new form of chemotherapy – was tried and the hospital noticed 'a brief flicker of response', but no more. After two weeks, unable to eat, too feeble to lift her head, sedated by morphine, she fell into delirium. She imagined that she was riding her bicycle across

Putney Common during an air raid. When Peggy came into the room, she said, 'I'm glad you've come, Aaron. Now you can help me with those graphs.'

On Wednesday 16 April, *The Times* carried an exhilarated report on the Brussels World's Fair, 'Atomic Crystal Gazing in Brussels', which emphasised the scientific thrust of the exposition. Rosalind never saw it. She died that afternoon.

The cause of death was given as bronchopneumonia, secondary carcinomatosis and carcinoma of the ovary. The brief entry on her death certificate said a great deal in a few words: 'A Research Scientist, Spinster, Daughter of Ellis Arthur Franklin, a Banker.'

When the news came, Ken Holmes and James Watt, her PhD student, wept. Finch crept into a nearby church. He could not believe it; he had seen her on her last day at the lab, walking by his door, dignified, head high as always. They all thought themselves tremendously lucky to have worked with her and even then were completing papers bearing her name, four of which would be published posthumously.

The funeral was held next day at the United Jewish Cemetery in Willesden. Vittorio Luzzati, who was in London, came. So did the team from the lab. The mourners held themselves in two distinct groups – family and scientists. There was little conversation and no gathering afterwards.

Rosalind was buried in the Franklin family plot, near her Franklin great-grandparents and her grandparents. Each of the gravestones of Arthur Ellis Franklin and his wife Caroline proclaims, 'This stone was obtained from the Holy Land.' Rosalind's says, 'Scientist: her work on viruses was of lasting benefit to mankind.'

Rosalind sank into her grave, as Yeats said of Keats, 'His senses and his heart unsatisfied.' Yet her work surpassed her most optimistic expectations. She had published, on her own or with others, thirty-seven scientific papers. Her death was noticed in *The Times, New York Times* (acknowledging her 'widespread recognition for her research on virus structure': or: 'one of a select band of pioneers unravelling the structure of nucleo-proteins in relation to virus diseases and genetics') and *Nature*.

J.D. Bernal's long admiration for Rosalind was expressed with erudite grace in a masterly obituary which appeared under a two-column heading in *The Times* on 19 April. He opened with the declaration 'Rosalind Franklin's early and tragic death is a great loss to science' and the statement that she had made a distinguished name in two very different branches of research, the study of coal and coke, then in nucleo-proteins and virus structure. Bernal then gave deserved attention to her earlier work:

> She discovered in a series of beautifully executed researches, the fundamental distinction between carbons that turned on heating into graphite and those that did not. Further, she related the difference to the chemical constitution of the molecules from which the carbon was made. She was already a recognised authority in industrial physicochemistry when she chose to abandon this work in favour of the far more difficult and more exciting fields of biophysics.

Then he moved on to her work on deoxyribonucleic acid (DNA) at King's College London:

> By the most ingenious experimental and mathematical techniques of X-ray analysis, she was able to verify and make more precise the illuminating hypothesis of Crick and Watson on the double spiral structure of this substance. She established definitely that the main sugar phosphate chain of nucleic acid lay on an outside spiral and not on an inner one, as had been authoritatively suggested.

When she moved to Birkbeck, Bernal wrote, she took up the tobacco mosaic virus, 'almost at once using the techniques she had already developed, made notable advances on it . . . She then made her greatest contribution in locating the infective element of the virus particle – its characteristic ribose nucleic acid.' He described her international recognition, particularly the American grant, and her new direction, the polio virus. He praised not only the 'apparently effortless skill' of her brilliant individual research but her gift as an organiser of research, 'as the small and devoted team she gathered round her bears witness'. 'Her life,' he concluded, 'is an example of single-minded devotion to scientific research.'

Bernal surpassed himself in his obituary for *Nature*: 'As a scientist Miss Franklin was distinguished by extreme clarity and perfection in everything she undertook. Her photographs are among the most beautiful X-ray photographs of any substance ever taken.'

Amid a long and considered summary of her work and her foreign collaboration and a tribute to her courage in working right up to the end, Bernal subtly addressed the apportionment of credit for the discovery of the double helix of DNA:

> In this close collaboration between the Cambridge and London schools it is difficult to disentangle all the contributions of individuals, but what Miss Franklin had to give was the technique of preparing and taking X-ray photographs of the two hydrated forms of deoxyribonucleic acid and by applying the methods of Patterson function analysis to show that the structure was best accounted for by a double spiral of nucleotides, in which the phosphorus atoms lay on the outside.

Her family was stunned by Bernal's obituary in *The Times*. Coinciding, as it did, with the exhibition of Rosalind's model at the Brussels World's Fair, it made them realise how very good a scientist she had been. As her Aunt Alice wrote to Anne Sayre a week after Rosalind's death, 'She was very modest and also knowing it was Greek to us – she did not waste time in talking about her work with those who would irritate her by silly questions.'

This self-deprecating comment from a highly able woman illustrates (as does, in many ways, Rosalind's entire career) the isolation of the scientist, cut off from ordinary discourse, even with loved ones, about what is of most intense interest in his or her daily life.

After Rosalind's death, Klug invited Muriel Franklin to visit 21 Torrington Place to show her where and how her daughter had worked. He showed her around the small top-floor room, with its neatly labelled test-tubes and the white lab coat still hanging on the back of the door. He praised the originality and delicacy Rosalind brought to her experiments.

His words drove the grieving mother back – perhaps to thoughts of the homesick little girl at boarding-school knitting a scarf for Nannie. 'Rosalind was always good at sewing,' she said.

EPILOGUE

Life After Death

'CONCERNING ROSALIND,' Maurice Wilkins wrote James Watson in 1966, 'is there any mention in your book that she died?'

'Your book' was 'Honest Jim' an early draft of *The Double Helix,* Watson's candid, fast-paced account of the discovery of the structure of DNA. In it Rosalind Franklin is 'Rosy', the termagant who hoarded data she couldn't comprehend, treated men like naughty little boys and wore dresses even dowdier than those of the average Englishwoman.

Watson could not have given the world his 'Rosy' if Rosalind had been alive. He began writing the book as a Harvard professor, three years after he, Francis Crick and Maurice Wilkins had won the Nobel prize for medicine and physiology. Watson submitted the book to Harvard University Press, which liked it but required the written consent of those prominently and candidly mentioned. Francis Crick and Wilkins objected most strongly – in Crick's case, with some anger. Linus Pauling also cast a veto. In a fierce letter to Crick, with nine copies, including one to Nathan Pusey, president of Harvard, Pauling condemned the portrayals of himself, his wife, his son, Francis Crick, Sir Lawrence Bragg and Rosalind Franklin.

Rosalind's brother Colin, as one of Rosalind's trustees and a publisher, was one of those sent a draft of the book. He was outraged. He shot off a cable containing what Watson considered 'a rather hysterical reaction' about 'defaming the dead'. When Charlotte Franklin, Colin's wife, followed with a 'sensible'

(Watson's word) letter, Watson considered that he should perhaps put in an epilogue.

Max Perutz too complained about the treatment of Rosalind. As he later wrote in the London *Daily Telegraph,* 'I was furious about his maligning that gifted girl who could not defend herself because she died of cancer in 1958; but I could not get him to change it . . . Not that she was unattractive or did not care about her looks. She dressed much more tastefully than the average Cambridge undergraduate . . .'

In response, Watson composed a pious epilogue stating what a fine scientist Rosalind Franklin had been, and how, as a young man, he had not appreciated the difficulties of a woman making her way in the man's world of science.

The many revisions Watson made to the early drafts did not placate his critics. Wilkins wrote T.J. Wilson, the director of Harvard University Press, to declare that although some of the grosser factual errors had been corrected, he nonetheless felt it disgraceful that a university press should be party to a distinguished member of its staff making such a display of immaturity and bad taste. Watson's book, Wilkins declared, was 'unfair to me, to Dr Crick and to almost everyone mentioned except Professor Watson himself'.

Faced with such formidable opposition, Harvard's Board of Overseers made its press drop the book. It complied, and lost a bestseller. Instead, *The Double Helix* was published by the Athenaeum Press in New York in February 1968, and by Weidenfeld and Nicolson in London in May, the original portrait of Rosy intact. The stilted afterword tacked on to the end of a racy tale did nothing to erase the impression of the virago ready to spring.

Rosalind's parents were deeply upset: bad enough to have lost their daughter, now this gratuitous and painful insult. Aaron Klug, visiting them, suggested that the book at least ensured that Rosalind would always be remembered. Muriel Franklin replied, 'I would rather she were forgotten than remembered in this way.'

* * *

The Double Helix was an instant success, welcomed for dispelling the myth that science is done by dispassionate intellects moving with deliberation towards defined goals. *Nature* called the book 'spellbinding and a considerable public service'. Jacob Bronowski said in *The Nation* that the book 'communicates the spirit of science as no formal account has ever done'. In the United States it was nominated for the National Book Award.

J.D. Bernal, reviewing *The Double Helix* for *Labour Monthly,* gently corrected the book's portrait of Rosalind while at the same time deflating Watson and Crick. He wrote 'a decisive break-through in human thought is not necessarily the work of an individual genius but only of a pack of bright and well-financed research workers following a good well-laid trail'. Bernal credited Rosalind with a significant part in laying the trail:

> For 'Rosy', in the book – I had come to know and respect her and to admire her too, as a very intelligent and brave woman who was the first to recognise and to measure the phosphorus atoms in the helix, which proves to be the outer one, thus showing Pauling to be wrong and the helix to be a double one, though this inference is not drawn.

Praise for the book was not unanimous. The 1960s were not the 1950s; the women's movement had begun. By 1968, lines such as: 'Clearly Rosy had to go or be put in her place . . . ; 'Certainly a bad way to go out into the foulness of a heavy, foggy November night was to be told by a woman to refrain from venturing an opinion about a subject for which you were not trained', and 'the best home for a feminist was in another person's lab' were as unacceptable as jokes about women drivers and dumb blondes.

Mary Ellmann, whose *Thinking About Women* appeared the same year and launched feminist literary criticism, recognised the misogyny underlying *The Double Helix*. In *The Yale Review* she mocked its pretence to show 'the scientist as human being, with genes in the morning, girls in the evening' and commented sarcastically, 'The only contradiction of this sensible balance is

Rosalind Franklin, the woman who *studies* DNA like a man . . . Why couldn't she content herself with playing assistant to Wilkins (and over his shoulder, to Crick and Watson)?' Elizabeth Janeway, in another feminist text, *Man's World, Woman's Place,* attacked as misogynist even Watson's portrait of Odile Crick as the air-headed pretty wife who thinks that gravity stops two miles up.

In ensuing decades, the myth of the wronged heroine has grown, nourished by the fact of Rosalind's early death. Rosalind Franklin has become the symbol of women's lowly position in the pantheon of science. In 1997, when an American neuroscientist, Candace Pert, was passed over for the Lasker Award (sometimes called 'the American Nobel') she felt she deserved, she blamed anti-female prejudice. It was just what had happened to Rosalind Franklin, she maintained. In *Molecules of Emotion,* Pert went so far as to suggest that Franklin's cancer 'had been exacerbated by the humiliation she suffered at the hands of these, and probably many other, old boys'.

The Double Helix is now established as a twentieth-century classic, published in eighteen languages: a candid young-man's-eye view of one of science's great discoveries. Watson wrote what, at twenty-three, he felt and saw happening. Yet a cloud of guilt hovers over the tale from the very first page. The book opens in 1955, with Jim Watson climbing a slope in the Swiss Alps. Suddenly he recognises one of the party of climbers coming down. It is Willy Seeds, from King's College London. Instead of pausing for a chat with an old friend in an unfamiliar place, Seeds merely says, 'How's Honest Jim?' and passes by.

The reader has no way of knowing that the ever-sardonic Seeds meant the very opposite of 'honest'. In 1955, only two years after the discovery of the double helix, scientists at King's had not forgiven Watson for helping himself to King's data to win fame. Seeds's gibe clearly hit home, for Watson made 'Honest Jim' the book's working title. Another discarded title – 'Base Pairs' – carries the same self-accusation.

Watson was intentionally ironic. He genuinely liked the sound of 'Honest Jim' for its echoes of *Lucky Jim* – Kingsley's Amis's 1954 comic masterpiece about a maladroit young instructor who exposes the pomposity of the British academic establishment. It may not be too far-fetched to think that in Watson's unconscious, as he shaped his narrative, lodged not only the bumbling honest Jim Dixon but the neurotic female lecturer Margaret Peel ('quite horribly well done', said the *New Statesman*), with her tasteless clothes and utter ignorance of how to appeal to a man.

Over the years, Watson repeatedly indulged in public admissions of unease. In 1999, in his book *A Passion for DNA*, he looked back to the publication of *The Double Helix* and joked: 'I daydreamed that the *New Yorker* might print it under the rubric "Annals of Crime" because there were those who thought Francis and I had no right to think about other people's data and had in fact stolen the double helix from Maurice Wilkins and Rosalind Franklin.'

As decades went by, Watson defended himself before young audiences who had little idea of what he was supposed to be guilty of. In London in 1984, addressing the Science Society at Rosalind's old school, St Paul's, he declared that he and Crick had not robbed Rosalind. 'We used her data to think about, not to steal,' he said. He tried to explain why he and Crick and not she had made the discovery: 'It *wasn't* because she was a woman or that we were more intelligent. We were more interested in DNA than she was, more interested because of our education and our friends.' He went on to make some remarks about Rosalind's inflexibility, then to inform the schoolgirls: 'She had terrible relations with her family. Indeed she went to stay with the Cricks after her hospital treatment.'

(Afterwards, the head of chemistry at St Paul's, Richard Walker, candidly recorded in his notes: 'quite the most provocative and stimulating lecture we've had! The science block is still humming to the comments of "that horrible man".')

Fifteen years later, and forty-six years after he walked into

Rosalind's room at King's and saw her bending over the lighted box, Watson was still justifying himself. As President of Cold Spring Harbor Laboratory, returning to Harvard to inaugurate its Center for Genomic Research, he began with a flashback to that critical day in January 1953:

> Let's just start with the Pauling thing. There's a myth which is, you know, that Francis and I basically stole the structure from the people at King's. I was shown Rosalind Franklin's x-ray photograph and, Whooo! that was a helix, and a month later we had the structure, and Wilkins should never have shown me the thing.
>
> I didn't go into the drawer and steal it, it was shown to me, and I was told the dimensions, a repeat of 34 Ångströms, so, you know, I knew roughly what it meant and, uh, but it was that the Franklin photograph was the key event. It *was*, psychologically, it mobilised us . . .

One new element added in the repeated retelling of the story is the admission that Rosalind's Photograph 51 was the pivotal moment in the discovery of the double helix.

Something was done that ought not to have been done, but what? Not the showing of Photograph 51. That concerned Raymond Gosling. He had participated in the work and, as has been said, he had every reason to give the small square X-ray print to Wilkins. If Wilkins was unwise in letting Watson have a look at it, there was no intended subversion; he did not, in that pre-photocopying era, make Watson a copy.

Rather, Watson's continuing sense of unease may well derive from the use of Rosalind's experimental data behind her back and *never telling her openly* – not even in the subsequent years of their friendly collaboration. Neither did Crick. Such acknowledgement as they gave her was very muted and always coupled with the name of Wilkins. For example a lengthy footnote appended to their 1954 paper for the *Proceedings of the Royal Society* says, well into the text,

> The information reported in this section [about the two different forms, A and B, of DNA] was very kindly reported to us prior to its publication by Drs. Wilkins and Franklin. We are most heavily indebted in this respect to the King's College Group, and we wish to point out that without this data the formulation of our picture would have been most unlikely, if not impossible.

The fact was that Rosalind did not report this information to them herself and she was not even speaking to Wilkins. To her dying, day Rosalind could not have dreamed that Watson and Crick would be declaring from public platforms half a century later that they could not have found the double helix in March 1953 without her experimental work.

Some scientists have accused Watson's book of undermining the ethics of science, demonstrating to young people that winning justifies all. Rosalind's friend David Harker, formerly of Brooklyn Polytechnic, later head of biophysics at the Roswell Park Memorial Institute at Buffalo, New York, put his objections thus:

> the real tragedy in this affair is the very shady behavior by a number of people, as well as a number of unfortunate accidents, which resulted in the transfer of information in an irregular way . . . I would never have consciously become involved in anything like this behavior . . . And I think these people are – to the extent that they did these things – outside scientific morals, as I know them.

Why did Watson create Rosy the Witch? A plausible hypothesis holds that the character was a rationalisation of Watson's guilt – a creature so hostile and uncooperative that there was no alternative to taking what you need by stealth. Gunther Stent, biochemist and editor of the Norton Critical Edition of *The Double Helix,* compares the book's moral dilemma to Lawrence Kohlberg's tale of 'Penniless Heinz and the Mean Druggist': a good husband steals from a mean druggist the medicine essential to save his wife's life.

From the feminine point of view, the wicked Rosy is a variant of an older myth, 'She asked for it', that traces back to Eve: the woman is guiltier than the male. Unwittingly, Nannie Griffiths drew on this ancient lie when blaming young Rosalind for complaining that Colin had hit her with a cricket bat: 'Well dear, you shouldn't have been teasing him.'

What cannot be denied is that 'Rosy' was essential as villainess for the plot for what Wilkins sometimes called 'Jim's novel'. Extraneous details, such as later friendship or early death, would have spoiled the narrative.

Unhappily, Rosalind's denigration did not end with *The Double Helix*. When outright mockery became impolitic, she began to be damned with faint praise; she has been called 'sound' and 'a good experimentalist', her ability and intelligence downgraded. It has been suggested that she was plodding, that she could not understand her own data or work in teams, accept criticism or use imagination.

That none of these alleged inadequacies manifested themselves in Rosalind's work on viruses or coal is ignored by her detractors. Instead, her failure to get the DNA structure in twenty-seven unhappy months at King's College is laid to an almost wilful blindness. Horace Freeland Judson, the science historian, has said that when, in 1952, Rosalind failed to listen to Crick's warnings about alternative explanations for her DNA data, 'Franklin betrayed a grievous slowness of intuitive response.' Yet the solution she reached in her classic papers on coal has been called 'quite witty', while her TMV work is acknowledged as outstanding. Marjorie M'Ewen, who worked with Rosalind at King's College, said, 'Rosalind's final, brilliant, work on TMV received such universal acclamation it is tragic that her scientific reputation was not allowed to rest thereon.'

It was not. The more sophisticated Crick is not blameless. In 'How to Live with a Golden Helix', an article in *The Sciences* in September 1979, he wrote:

> Rosalind's difficulties and failures were mainly of her own making. Underneath her brisk manner she was oversensitive and, ironically, too determined to be scientifically sound and to avoid shortcuts. She was rather too set on succeeding all by herself and rather too stubborn to accept advice easily from others when it ran counter to her own ideas.

But she was not 'set on succeeding all by herself'. Her close collaboration with Mering and later Klug, Holmes and Finch shows that she could be an inspiring teamworker. Her letters to Crick himself demonstrate that she not only invited but welcomed his tough criticisms on her TMV papers, even when these contradicted her own ideas.

Like Watson, Crick felt the need to find a cause for her obstinacy and found it in her family. In 'How to Live with a Golden Helix', he said:

> The major opposition Rosalind Franklin had to cope with was not from her scientific colleagues, nor even from King's College, London (an Anglican foundation, it should be noted, and therefore inherently biased against women), but from her affluent, educated and sympathetic family who felt that scientific research was not the proper thing for a normal girl.

Crick did not know Rosalind's family. He did not know they were Jewish – certainly a major element in her feeling that King's was 'inherently biased' against her. In reality, Ellis Franklin never opposed his daughter's career. There would have been no point. He and his wife did not try to stop her climbing dangerous peaks, moving to France, hitchhiking alone through Israel: Rosalind did what she wanted to do. However, she could see for herself that she was the odd one out in a family of intermarrying cousins and unsalaried wives.

Crick was nearer the mark when telling Anne Sayre in 1970 in a taped interview that Rosalind, as he came to know her through her TMV research, had 'a good, hard, analytical mind, really first-class', but that she lacked intuition – 'Or mistrusted it. Perhaps mistrusted it.'

Rosalind *did* mistrust intuition, with a wariness for which her Jewishness is as relevant as her gender. In *The Cousinhood: The Anglo-Jewish Gentry* published in 1971 not long after *The Double Helix*, its author Chaim Bermant writes:

> A Jew often feels compelled to try that much harder than his colleagues; a woman in a man's world has a similar compulsion. Rosalind perhaps tried too hard on both scores and approached her work with a jealous determination which some of her colleagues found alarming. She seemed to carry a constant air of embattlement about her, and felt that her first-class ability and achievements were not given due recognition.

Blame the victim. The belief that Rosalind brought her fate on herself extends even to her cancer. It is often suggested that careless exposure to radiation was the cause of her disease. However, such anecdotal evidence as there is of her taking exceptional risks is counter-balanced by praise of her extreme caution as a scientist, also by the widely repeated observation that all laboratory staff in the 1950s did things that, in Ken Holmes's words 'today would bring down the wrath of the safety inspectorate'.

It may be relevant that others in the Franklin family have suffered from cancer. Possibly they fell victim to the 'Ashkenazi gene' – two genes actually, BC1 and BC2, found to be responsible for the disproportionate percentage of cancer deaths among Jews of northern European descent, with a particularly high incidence of female breast cancer.

For all the retributive gestures, Rosalind continues to be overlooked in accounts of the discovery of the double helix. In Bryan Sykes's *Seven Daughters of Eve,* published in 2001, about the genetic markers of European inheritance, Sykes describes how Watson and Crick worked out the molecular structure of DNA:

> This entailed making long crystalline fibres of DNA and bombarding them with X-rays . . . The deflected X-rays made a regular pattern of spots on the film . . . After many weeks spent building different models . . . Watson and Crick suddenly found one which fitted exactly with the X-ray pattern.

Whose X-ray pattern? Long crystalline fibres made by whom? Bombarded by whom? Film owned by whom? Rosalind seems doomed to remain the invisible woman in many minds, the faceless nurse who hands the surgeon the scalpel.

The air of injustice hanging over the name of Rosalind Franklin has inspired some belated acts of contrition. Immediately following her death, Don Caspar, Klug and Bernal led efforts to organise a memorial fund in Rosalind's name. That never got off the ground. Instead, a registered charity, the Rosalind Franklin Bequest, was created out of the residue of her estate, about £5,000, which gave small grants to deserving applicants.*

That apart, silence followed for many years until Anne Sayre's combative biography of Franklin in 1974 took aim at the Watson caricature. Awareness was aroused, but slowly. In 1984, St Paul's Girls' School recognised her as one of its most illustrious pupils. The Rosalind Franklin Design, Technology and Engineering Workshop was opened at a ceremony attended by her family and former colleagues. (At the event Holmes, who came over from Heidelberg for the occasion, ran into Gosling, then professor of medical physics at Guy's and St Thomas's Hospitals, and said, 'So you were in love with her too?' – a light-hearted remark that did not conceal Holmes's unhealed grief. As he wrote to Anne Sayre in 1971, he was unable to help with her book: 'The circumstances of her death and her bravery during her illness were such that my feelings about Rosalind are still very intense.')

As her posthumous reputation grew, Rosalind was accorded

* In 2001 Colin Franklin and Jenifer Franklin Glynn gave the capital, by then about £150,000, to Newnham College for its bursary fund.

a highly sympathetic portrayal in 1987 in the BBC *Horizon* programme's dramatisation, 'Life Story', with Juliet Stevenson and Alan Howard as the ill-matched Franklin and Wilkins, Jeff Goldblum as Watson and Tim Piggott-Smith as Crick. (Well-researched, its scientific details are impeccable, and J.T. Randall wears exactly the kind of bow tie he sported in the early 1950s. The film, nonetheless, to Aaron Klug made Rosalind 'nun-like', whereas the Rosalind he knew was vivacious, opinionated and fun-loving. Anne Sayre too, who found Rosalind a sparkling and amusing friend, disliked the film's 'droopy drudge . . . sitting up half the night with her Beevers-Lipson strips'. Anne remembered well that Rosalind did do hundreds of hours of hand calculations, but 'with concentration and burning enthusiasm . . . she *did* have real insight, she *did* know where she was headed, she *did* have unusual capacity to interpret small clues'. The film that needed to be made, said *New Statesman*, 'is of a brilliant woman trying to make her way in a man's world, having her work used behind her back, and finally being misrepresented in a book published when she was no longer alive to reply'.

Honours continue to accumulate. In 1998 the National Portrait Gallery hung Rosalind's photograph beside that of Wilkins and beneath those of Watson and Crick. In 1992 English Heritage placed a blue plaque on the dark-red mansion block of Donovan Court, Drayton Gardens, South Kensington. A Heritage spokesman said, 'Franklin never received adequate recognition. We are particularly anxious to commemorate important women as 90 per cent of the 600 plaques put up in the last 125 years are to men.' In 1995, sixty-seven years after failing to honour her with an obituary in the Newnham Roll, Newnham College dedicated a graduate residence in her name and placed a bust of Rosalind Franklin in its garden.

At the Science Museum in South Kensington, the TMV model shown in the Science Pavilion at the Brussels World's Fair was acquired for exhibition and kept there until 1964 when Max Perutz took it to Cambridge.

As if to outdo all the rest, at the beginning of the new millennium, King's College London, in a genuflection bordering on political correctness, honoured her in March 2000 with the dedication of its new Franklin-Wilkins Building, opened by the Chancellor of London University, the Princess Royal. During the ceremony, Francis Crick appeared on video, and Watson came in person, to say that her contribution was critical to their discovery. The building thus links the names of a professor who had worked and taught there for fifty-three years with the name of an associate worker who had left (or been pushed out) after little more than two years and rarely spoken of with any warmth within its walls before the name of the refurbished warehouse on the South Bank was chosen.

'*Waarom kreeg "Rosy" geen Nobelprijs?*' The question posed by the Dutch newspaper *Vrij Nederland* on 15 August 1998 will not go away, even if the answer is simple. 'Rosy' never won the Nobel prize because, when Watson, Crick and Wilkins got theirs in 1962, she was dead. The prize is never awarded posthumously. The inevitable rephrasing – 'If Rosalind had lived, would she have won the Nobel prize?' – is meaningless and belongs in the Alternative Futures file along with 'What if Kennedy had not gone to Dallas?'

When the question is rephrased to an assertion, 'They would have had to give it to her rather than to Wilkins had she lived,' the rejoinder is easy. Not at all. 'They' are not compelled to do anything except respect the Nobel rules, which do not allow more than three people to share a prize in any single category. The Nobel committee in Stockholm chooses the winners from the recommendations sent in from each country by, among others, its national laureates – a system which encourages fraternity.

Indeed, when Wilkins's name was announced as a co-winner with Watson and Crick in 1962, Sir Lawrence Bragg wrote Max Perutz (who won the Nobel for physics that same year) of his 'great joy' that Wilkins had been linked with the other two. 'It will undo the bitterness he [Wilkins] felt when Crick and Watson proposed their structure,' Bragg wrote. 'I pressed strongly for it in my recommendation.' Bragg was very influential in determining British nominations. It is hard to believe that, nudged possibly by J.T. Randall, Bragg would not have put Wilkins's name forward instead of Rosalind's. Wilkins was senior to Rosalind, a Fellow of the Royal Society, and had done the serious DNA preliminary work at King's that led to hers.

Speculation is hardly required to imagine the course Rosalind's life would have taken had she been less unlucky. Her long struggle to get her Virus Research Project on a permanent footing succeeded in 1960 when, with the Medical Research Council taking over from the US Public Health grant as backer, Klug, as head of the group, moved with Holmes and Finch to the new MRC Laboratory of Molecular Biology at Cambridge. There Klug and Finch remained for the rest of their careers. Holmes, the one who most directly continued Rosalind's work on the structure of TMV, moved in 1968 to the Max Planck Institute for Medical Research in Heidelberg and carried on his research from there.

Nobel statistics do not favour the female. In 1956 when Dorothy Hodgkin FRS solved the structure of Vitamin B12, the prize for chemistry that year was divided between the head of her department and a Russian chemist. For year after year after that, knowing she had been nominated several times, Hodgkin waited for the telephone to ring. It had not rung by 1962 when Max Perutz received his prize, embarrassed (he said later) to have his prize before she had hers. He used his Nobel credentials to lobby for her. At last, with the special honour of being a sole winner, Hodgkin received the prize for chemistry in 1964, the fifth woman scientist ever to win in the six decades of the prize. The attendant publicity called her an 'affable-looking housewife' and 'mother of three'. At the seventy-fifth anniversary of the Nobel prize in 1975, she was the only representative of fourteen women laureates from all subjects, including peace and literature. She was used to being the only woman at gatherings, she said. 'After all, women have come rather late to science.'

When the centenary of the Nobel prize was celebrated in 2001, Hodgkin remained the only British woman scientist to have won it. A conspicuous absentee from the Nobel pantheon of science is Jocelyn Bell, the Cambridge astronomer who discovered pulsars – energy emissions constituting a new class of stars. She was a graduate assistant at Cambridge when she made her discovery. The Nobel prize for physics in 1974, however, honoured her professor, Antony Hewish, for 'recognising the meaning' (in the words of the *Encyclopaedia Britannica)* of his assistant's observations.

Rosalind's name *was* praised from the Nobel platform in Stockholm but not by Watson, Crick or Wilkins. When they gave their Nobel addresses in 1962, only Wilkins uttered her name at all, mentioning her and Alex Stokes as two people at King's College London who 'made very valuable contributions to the X-ray analysis'. Sometime later, Randall (by then Sir John) wrote Gosling, 'I have always felt that Maurice's Nobel lecture did rather less than justice to this setting [Randall's biophysics lab at King's] and particularly to the contribution of yourself and Rosalind.'

In 1982, however, when Aaron Klug received the prize for chemistry, he spoke movingly of his late colleague. Rosalind Franklin, he said, had introduced him to the study of viruses and set an example of tackling large and difficult problems: 'Had her life not been cut tragically short, she might well have stood in this place on an earlier occasion.'

Klug has been Rosalind's staunchest defender. Several months following the publication of *The Double Helix,* he wrote a long article for *Nature,* saying that Watson did not pretend to tell more than one side of the story. As 'her last and perhaps closest scientific colleague', he said, he had carefully studied her laboratory notebooks and found that, far from being anti-helical (as Watson, Crick and Wilkins continued to maintain) she had set out the evidence for a helical structure for DNA as early as her Turner and Newall report in February 1952. Although she had retreated from this position for the A form because of some of her subsequent findings, by the February and March of 1953, she was far nearer to solving the structure than anyone had realised:

> She would have solved it, but it would have come out in stages. For the feminists, however, she has become a doomed heroine, and they have seized upon her as an icon, which is not, of course, her fault. Rosalind was not a feminist in the ordinary sense, but she was determined to be treated equally just like anybody else.

Klug's was a robust defence, strengthened in 1974 by a startling discovery, which he also reported in *Nature*, of a missing manuscript dated 17 March 1953, which showed that Rosalind was even closer to coming upon the truth of the double helix than even he had realised. Tactfully, however, perhaps out of respect for his fellow FRSs, Klug did not mention how irregularly her data had been obtained, and how inadequate had been her formal acknowledgement, in 1953 and 1954, and in the 1962 Nobel addresses of the DNA trio.

Nor did Klug declare his personal interest in her defence. Klug owed Rosalind a debt of honour. By making him her principal beneficiary, she changed his life, made it possible for him and his wife to buy a house and to stay in Britain where he rose to great heights. He became successively Sir Aaron, winner of the Nobel prize, holder of the Order of Merit (a tribute to greatness, in the personal gift of the Queen) and, from 1995 to 2000, President of the Royal Society.

Some criticism belongs to the Nobel prize itself. Founded in 1901 it is the world's most coveted intellectual award, but it is also arbitrary, inherently unfair and possibly damaging. Many who deserved it have not got it; many who didn't, have. The literature prize offers glaring examples: Thomas Mann, Leo Tolstoy and James Joyce were passed over, Somerset Maugham and Pearl Buck honoured. In science the effect is particularly baleful, because science is collegial. As research becomes progressively more expensive, moreover, scientific discoveries are harder to attribute to particular people. The Nobel prize, by canonising individuals, disguises the truth that they are all, in Newton's famous phrase, standing 'on giants' shoulders' and on each other's as well.

The list of deserving scientists who never got the summons to Stockholm is long, glittering and bitter, and contains a number of women: the physicist Lise Meitner and the Cambridge astronomer Jocelyn Bell. Oswald Avery, who discovered that DNA was the genetic material, was deprived of the prize by the persistent and unfair criticism of his Rockefeller colleague Alfred Mirsky. In the view of Erwin Chargaff, whose finding of the base pairs of DNA went uncelebrated, 'Avery should have gotten two Nobel prizes for his discovery. I have no complaint when I think of Avery.'

Such omissions and exclusions would not matter except that the Nobel prize changes lives, and divides colleague from colleague by touching the winners with a magic of which the generous prize money is only part.

Rosalind Franklin did not have her eyes on the prize. Nor did she worry about having been outrun in a race that no one but Watson and Crick knew was a race. She died proud of her world reputation both in coal studies and in virus research, and of her list of published papers that would do credit to any scientific career, let alone one that ended at the age of thirty-seven. Had she not gone into science, the early 1950s would have had little understanding of what kinds of coal make graphite, slower knowledge of the A and B forms of DNA and of the characteristics of TMV. There would have been a delay in getting to the DNA structure and in the revolution that followed. The careers of her collaborators and of the Nobel trio who benefited from her data could have fallen far short of the heights they reached.

Rosalind knew her worth. With every prospect of going on to further significant achievement and, possibly, personal happiness, she was cheated of the only thing she really wanted: the chance to complete her work. The lost prize was life.

NOTES

References to quotations and other cited passages are indicated by the first words of the relevant text or, in a few cases, by the key words.

The seminal papers in *Nature* on 25 April 1953 by Watson and Crick, Wilkins, Stokes and Wilson, and Franklin and Gosling can also be found in the Critical Edition of *The Double Helix*, as can Aaron Klug's analysis of Rosalind's notebooks, in *Nature*, in 1968 and 1974. Information on the location of specific collections of papers appears in the Acknowledgements. Letters and memoirs for which no location is given are in Franklin family hands. Where a letter is undated it is identified by its heading, which may include part of a date and/or the address.

The following abbreviations are used frequently in the Notes:

Archives and Private Collections

ARC: Agricultural Research Council
ASA: Anne Sayre Archives, American Society for Microbiology Archives, University of Maryland, Baltimore County
CAC: Churchill Archives Centre, Churchill College, Cambridge University
JNC: Jeremy Norman Collection of Molecular Biology, Novato, California
PA: Linus and Ava Helen Pauling Collection, Oregon State University
PRO: Her Majesty's Public Record Office, Kew
RS: The Royal Society, London

People

AK: Aaron Klug
AS: Anne Sayre
AW: Adrienne Weill
CF: Colin Franklin
DC: Don Caspar
EF: Ellis Franklin
FHCC: Francis Crick
GCD: Gertrude Clark Dyche
HFJ: Horace Freeland Judson
JC: Jane Callander
JDB: John Desmond Bernal
JDW: James D. Watson
JG: Jenifer (Franklin) Glynn
JTR: John T. Randall
KCH: Kenneth C. Holmes
LP: Linus Pauling
MF: Muriel Franklin
MP: Max Perutz
MW: Maurice Wilkins
PP: Peter Pauling
RF: Rosalind Franklin
RG: Raymond Gosling

UR: Ursula Richley
VL: Vittorio Luzzati
WLB: William Lawrence Bragg

PROLOGUE

'It has not escaped': J.D. Watson and F.H.C. Crick, 'A Structure for Deoxyribonucleic Acid', p. 737.
'Our dark lady': MW to FHCC, 'Saturday', 7 Mar. 1953, in Olby, *The Path to the Double Helix*, p. 414.
'From the evidence': A. Klug, 'Rosalind Franklin and the Discovery of the Structure of DNA', pp. 808–10, 843–4. See also, Crick, in A. Sayre, *Rosalind Franklin & DNA*, p. 214.
'As a scientist': J.D. Bernal, 'Obituary, Rosalind Franklin', *Nature*.

PART ONE

ONE *Once in Royal David's City*

'known as The Cousinhood': C. Bermant *The Cousinhood*, p. 1 and S. Brook, *The Club*, p. 33.
'Benjamin Wolf Franklin lived in the City': A.E. Franklin, *Records of the Franklin Family and Collaterals*.
'Keyser's, a source of employment': Bermant, op. cit., p. 282.
'bought the house of George Routledge': ibid.; author's interview with Norman Franklin; CF to author, 5 Sep. 2001.
'Rosalind's parents': M. Franklin, *Portrait of Ellis*, p. 63.
'The whole idea': EF to CC, 24 Dec. 1944, ibid., p. 267.
'It may appear strange': A.E. Franklin, *Records*, p. 4.
'the Franklins, as he saw them': 'The idea that we were descended from King David was always a point of family badinage', CF to author, 8 Oct. 2001.
'a fair proportion': A.E. Franklin, op. cit., p. x.
'Handed over': the memo dated 30 June 1920, is reproduced in H.F. Bentwich, *If I Forget Thee*, p. 170, and the event well described in T. Segev, *One Palestine, Complete*, p. 155 and p. 155n., who says that the light-hearted memo was Bols's joking revenge for Samuel's parody of his appointments list on 1 Apr.; also, that Samuel always insisted that the receipt was a joke, not an administrative document. It was sold at an auction in New York many years later for $5,000.
'granted self-government': Segev, op. cit., pp. 34–5. The actual memo, 'The Future of Palestine', 21 Jan. 1915, is in the Cabinet Office papers at the PRO, CAB 37/123/43.
'The Jewish brain': ibid., p. 35.
'which may prejudice': ibid., pp. 34–5.
'Somehow Englishmen': from a book of parliamentary sketches published in 1871, in 'Simon Hoggart's Diary', *Guardian,* Review, 30 Sep. 2000.
'a curious illustration': H.H. Asquith to Venetia Stanley, 28 Jan., 1915, in B. Wasserstein, *Herbert Samuel*, p. 210.
'in this reign': C. Dickens, *A Child's History*, p. 151.
'not our kind of Jew': H. Cooper and P. Morrison, *A Sense of Belonging*, pp. 71–4, 78, 82.
'the hidden discourse of the Jews': S. Gilmer, p. 214.
'Intermarriage': ibid., p. 258 and pp. 288–9; Charcot, Freud's teacher, in 1888 published the view that inbreeding was the cause of the

higher incidence of insanity among Jews.

'Jewish heart': W. Shakespeare, *The Merchant of Venice*, Act iv. i.

'raven-tressed': Sir W. Scott, *Ivanhoe*, pp. 92, 99.

'The Jew is everywhere': J. Buchan, *The Thirty-Nine Steps*, p. 18.

'a very charming': Rebecca West to Letitia Fairfield, 3 Apr. 1927, B. Kime Scott (ed.), *Selected Letters*, p. 91.

'How odd/of God' was composed by the British journalist William Norman Ewer (1885–1976).

'assumption of false names': H. Belloc in A. Julius, 'England's Gifts to Jew Hatred', pp. 11–14.

'Rooting out the Jew': ibid.

'Jews succeed against the grain': ibid.

'so very outspokenly': J. Glynn, *Tidings from Zion*, p. 42.

'a young man with long yellow hair': ibid., p. 52.

'I'm sure we don't': ibid., p. 54.

'unusual course . . . I trust': P. Fletcher-Jones, *The Jews of Britain*, p. 145. Although the House cheered Salomons, having repeatedly voted to admit Jewish members only to be overruled by the conservative House of Lords, he was removed as a trespasser. Not until 1858 did the Jewish Relief Act open Parliament to Jews by allowing them to omit the words 'the true Faith of a Christian' from the oath. Then Salomons, who in the interval become the first Jewish Lord Mayor of London, and Baron Lionel de Rothschild, who had been elected several times since 1847, took their seats.

'I like Europe': RF to parents, 29 Oct. 1950.

TWO *'Alarmingly Clever'*

'Ellis would lead': details of Ellis's background, war experience and family life are from M. Franklin, *Portrait*.

'we lived like': L.H.L. Cohen to author, 8 Dec. 1999.

'We are enjoying', Helen Bentwich to Norman Bentwich, 22 Aug. 1926.

'In 1926 British women had enjoyed the vote', if they were over thirty. The vote had been extended to women over thirty in 1918 when it was granted to men over twenty-one; women did not achieve equality with men in voting age until 1928.

'their mother, Caroline Jacob Franklin': family details in J. Glynn, 'Rosalind Franklin 1920–1958'.

'Ellis himself would not allow women employees': M. Franklin, op. cit., p. 189.

'Norland Place on Holland Park Avenue': Joane Keene (ed.), 'Norland Place School'.

'Muriel Franklin was a gentle': 'Obituary, Muriel Franklin', the Working Men's College *Journal*, Winter 1975–76; M. Franklin's privately circulated memoir, 'In the beginning'; JG to author, 14 Oct. 2001.

'I got two stars at school': RF to Caroline Franklin, 5 Mar. 1927.

'the five Ellis children': author's interview with UR, 20 Jan. 1999.

'It is Nannie's birthday': RF to Caroline Franklin, 10 Mar.

'When is Nannie': MF to AS, 11 Nov. 1974, Box 2, ASA.

'Do something!': C. Franklin, 'Vignettes of Rosalind', p. 1.

'One recalls a smug': M. Franklin, 'Rosalind', pp. 3–4.

'Rosalind enjoyed this trick of teasing': ibid., p. 4.

'A child being Lifted Up': drawing in RF's personal papers.
'Keyser's total profits': Roland Franklin, 'Ellis in the City', in M. Franklin, *Portrait*, p. 188.
'Dear Grandma': RF to Caroline Franklin from Hôtel Splendide, Marseilles. The game described was 'Kim's Game' taken from the Boy Scouts – a development skill from the Boer War.
'They are having a bazaar': RF to Caroline Franklin, 10 Mar.
'men of the working classes': M. Franklin, *Portrait*, p. 162n.
'volunteered to dust the books': Working Men's College *Journal*, 541, 1985, p. 17.
'to give the working man': ibid., 323, 1923, pp 161–6.
'ribbings': M. Franklin, 'Rosalind', p. 10.
'Please tell Roland': RF to parents, 11 May ?1930.
'She is obviously very happy': MF to Arthur Franklin, Bexhill-on-Sea, 'Sunday'.
'What is the new kitten like?': RF to parents, 'Sunday'.
'A half-timbered': 'Town's debt to Independent schools' place in Bexhill', *Bexhill Observer*, 4 Feb. 1967.
'There was a lantern lecture': RF to parents, 30 Nov. 1930.
'As they did not': RF to parents, 5 Oct. 1930.
'There was a very interesting lecture last night': ibid.
'Somebody must have made some mistake': RF to parents, 'Tuesday'.
'Please do not say anything': ibid.
'delicate': Glynn, 'Rosalind Franklin 1920–1958', p. 269.

THREE *Once a Paulina*

'At St Paul's': reported in the *Graphic*, 18 Feb. 1922, in H. Bailes, *Once a Paulina*, p. 87, whose title is borrowed for this chapter.
'It was always': author's interview with UR, 19 Jan. 1999.
'Rosalind's menstrual periods': R. Franklin medical history, University College Hospital Case No. AD 1651.
'I was told': RF to parents, 15 May.
'I am *very* cross': RF to parents, 26 May.
'I only got B': RF to parents, 'Thursday'.
'stodge': ibid.
'the only real disadvantage': H. Bentwich, *If I Forget Thee*, p. 7.
'Rosalind enjoyed stirring him up': M. Franklin, 'Rosalind', p. 4.
'the pivot of his emotional life': Dennis Overbye, *Einstein in Love*, p. 337.
'akin to that': ibid., p. 336.
'magnificent new block': this and the following material is taken from the 'Report of the Council on the Educational Work of St Paul's School for Girls', Matriculation and School Examinations Council, University of London, Feb. 1935.
'No one asked either of them to dance': Jean Kerlogue, 'Memories of Rosalind 1933–1938', privately circulated booklet.
'incredible innocence': author's interview with UR, 21 Jan. 1999.
'The weakness and lack of moral fibre': M. Franklin, *Portrait*, p. 180.
'We have just arrived home', RF to A.E. Franklin, 12 May 1937.
'we enjoyed plenty of fun': C. Franklin, Vignettes of Rosalind.
'nearly all north country': RF to parents, 16 Mar. 1938.
'reasonable chance of an award': RF to A.E. Franklin, 22 Oct. 1938.

'I am very much looking forward': ibid.

'whole or a half': author's interview with UR, 20 Jan. 1999.

'French mannerisms': RF to parents, 4 Jul. 1938, ASA.

'"masses" of sewing': RF to parents, 25 Jul. 1938.

'some of my own money': RF to parents, 4 Jul. 1938.

'You will, I know': Ethel Strudwick to Ellis and Muriel Franklin, 15 Jul. 1938.

'Rosalind Franklin showed great promise': Minutes of the Governors' Meeting, 15 Jul. 1938, St Paul's Girls' School archives.

'the most unconstitutional act of the century': Andrew Roberts, speaking on Channel 4, 19 Jul. 2000.

'Does this really mean': RF to parents, 29 Jul. 1938.

FOUR *Never Surrender*

'The myth has got': see M. B. Ogilvie, with K. L. Meek, *Women and Science*, p. 134 and Gaby Himsliff, 'Crusade targets sexism in science', *Observer*, 20 Jan. 2002.

'this is the first occasion': EF to Colin, 20 Oct. 1938, in M. Franklin, *Portrait*, p. 259.

'awful crowd of people': RF to parents, 11 Oct. 1938, in ibid.

'ridiculous fuss': RF to parents, 1 Feb. 1939, in ibid.

'happened to know': RF to parents, 20 Oct. 1938, in ibid.

'tell me all': RF to parents, 16 Mar. 1939, in ibid.

'Who and what': RF to parents, 5 Feb. 1939. The relative and family friend in question was Dr Redcliffe N. Salaman, FRS (1874–1955), an authority on the potato, from the genetic, sociological and economic point of view.

RF to parents, 20 Oct. 1938.

'most exciting': RF to parents, 15 Nov. 1938.

'over which Prof. Bragg': RF to parents, 2 Nov. 1938.

'What is a crystal?': RF undergraduate notebook labelled Minerology, 7/8, CAC.

'people who choose to go into science': JC interview with Stephen Bragg, 11 Jan. 1985.

'The complaint about my chemistry': RF to parents, 20 Jan. 1939.

can't find 'only be because he was American': ibid.

'Why are you so surprised': RF to parents, 'Wednesday'.

'I don't know what you see in Ros': GCD (the former Peggy Clark) to AS, 9 Jun. 1976, ASA.

'I would not have gone': RF to parents, 15 Nov. 1938.

'Apart from your letters': RF to parents, 20 Nov. 1938.

'The young American historian': A.M. Schlesinger Jr, *A Life in the 20th Century*, p. 208.

'unsuitable either climatically': *The Times*, 22 Nov. 1938, 'Jewish Refugees, Settlement Plans in the Colonies', 'Admissions to Britain Limited'.

'The statement effectively limited': S. Sofer, *Zionism and the Foundations of Israeli Diplomacy*, p. 34.

'I cannot see why there has not been more criticism': RF to parents, 24 Nov. 1938, in Schlesinger, op. cit., p. 207.

'Whereas my forefathers': Clause 34, will of A.E. Franklin, signed 16 Jul. 1935.

'I earnestly request my named': Clause 35, ibid.

'she knows her work': RF letter to parents, 18 Apr. 1939.

'I have made a frightful mess': RF letter to parents, 20 May 1939.
'if you are less intelligent': RF to parents, 'Tuesday', May ?1940.
'Now I really feel': RF letter to parents, 'Newnham Tuesday'.
'This holiday', described in M. Franklin, *Portrait*, pp. 200–1.
'Her father at first': Glynn, op. cit., p. 270, and JG to author, 2 Jul. 2001.
'I have just had a great triumph': RF to parents, 5 Oct. 1940.
'Geometrical basis': RF notebook headed 'X-ray Crystallography II', 7/3.39, CAC.
'white things looking like eggs': RF to parents, 13 Mar. ?1940.
'One of the social features': Gillian Sutherland, 'Nasty Forward Minxes', in S. J. Ormrod (ed.), *Cambridge Contributions*, p. 98.
'I am *not* one of the people': RF to parents, 26 Feb. 1940.
'that the reply': ibid.
'I don't understand': RF to EF, 16 May 1940.
'our King': RF to parents, 'Tuesday', May 1940.
'we are being beaten': RF to EF, 16 May 1940.
'Exams begin': RF letter to parents, 'Tuesday', May 1940.
'got a scrap of brain': RF to parents, ibid.
'She had got a first': information from RF's letter to parents, 12 Jul. 1940, and Dainton's to AS, 8 Nov. 1976, ASA.
'quite exceptionally bad': RF to parents, 'Tuesday', May 1940.
'You frequently state': RF to EF, no date; Glynn, op. cit., p. 272, places it as 'possibly summer 1940'.
'consciously a Jew': Glynn, ibid., p. 272.
'Ursula felt': author's interview with UR, 21 Jan. 1999.
'If it does not go': RF to parents, 12 Oct. 1940.
'I suppose "unfurnished"': RF to MF, 1940.
'I want to have you': Fred Dainton to AS, 8 Nov. 1976; subsequent details also from this long letter and its follow-up to AS, 24 Nov. 1976, ASA.
Rosalind's St Paul's friend: Jean Kerlogue: 'Memories of Rosalind'.
'bribed her fourteen-year-old brother': C. Franklin, op. cit.
'please send my evening dress': RF to parents, 18 Feb. 1940.
'I can't think why': RF to parents, 19 Nov. 1940.
'It was all rather exciting': RF to parents, 25 Nov. 1940.
'The general attitude': 'Women's Place in Industry, War-time Needs and After, Engineer Trainees', *The Times*, 16 Jan. 1941.
'She was inflexible': Dainton, op. cit.
'almost incapable': RF to parents, Newnham, 24 May 1940.
'I'm sure it sounds silly': RF to parents, 24 May 1941.
'not sleeping': ibid.
'Rosalind confessed': I.F. Neuner to AS, 22 Jan. 1971, ASM.

FIVE *Holes in Coal*

'bad tempered': this and other details of Norrish's character taken from Fred Dainton to AS, 24 Nov. 1976, ASA.
'on the pretext': RF to parents, 3 Oct. 1941.
'I've never had so much time': RF to parents, '7 Mill Road, Sunday'.
'a baby wireless': RF to parents, 30 Jul. 1941.
'50 per cent': A. Briggs, *The War of Words*, p. 48.

'one bomb in three': R. Neilands, *The Bomber War*, p. 58.
'You can think it': JDB to C.P. Snow, 11 Apr. 1961, Dept. of Physics Archive, Birkbeck College, University of London.
'Considering the experiences': RF to parents, 'The Lab, Wednesday'.
'I like long sentences': RF to parents, '7 Mill Road, Saturday'.
'quite incapable': RF to parents, '7 Mill Road, Tuesday'.
'When I stood up to him': RF to parents, 'The Lab, Wednesday'.
'he on the right': this and subsequent quotes about the dinner from K.C. Paice to AS, 14 Apr. 1976, ASA.
'not thrilling': RF to parents, '7 Mill Road, Sunday'.
'I don't know whether': ibid.
'I certainly don't mind': RF to EF, 1 Jun. 1942.
'I am extremely sorry': ibid.
'a room full of junk': ibid.
'she was extremely kind': author's interview with Marianne Weill Baruch, 11 Oct. 1999.
'as I can never see it': RF to parents, 27 Jul. 1942.
'I could hardly keep': RF to MF, 7 Jun. 1942.
'In industry': ibid.
'and a daily supply of liquid air': RF to parents, 27 Jul. 1942.
'I think the work sounds': ibid.
'Even washing-up': RF to Evi Ellis, 25 Jul. 1944.
'international reputation': see Peter J.F. Harris, 'Rosalind Franklin's work on coal, carbon and graphite', pp. 204–9.
'turned the signs around': Sayre, op. cit., p. 65.
'too good': Irene Neuner to AS, 22 Feb. 1971, ASA.
'driven more by my fear': A. Piper, 'Light on a dark lady', p. 152.
'There are only two': Kerlogue, op. cit.
'She wanted a swim': ibid.
'Irene will be taking': EF to CF, 10 Oct. 1943, in *Portrait*, p. 261.
'the house will be': ibid.
'Of course, for anyone': RF to Evi Ellis, 25 Jul. 1944.
'alien Jews': RF to CF, 24 Dec. 1944, *Portrait*, p. 267.
'Finding a queue': RF to parents, 4 Jul. 1945.
'He's merely expressed': RF to parents, 27 Jun. 1945.
'Forty-three years': Hertha Ayrton in Joan Mason, 'The Women Fellows' Jubilee', p. 127.
'Some press reports': R.L. Sime, *Lise Meitner*, p. 327.
'If ever you hear': RF to AW, 3 Feb. 1946, in Glynn, op. cit., p. 276.
'Thermal Expansion': D.H. Bangham and R. E. Franklin, 'Thermal Expansion of Coals and Carbonised Coals'.
'I make you notice': Gaetan Foquet to RF, 18 Jul. 1946.
'I am quite sure': RF to MF.
'She did not talk': Kerlogue, op. cit.
'I think she would have liked': ibid.
'abrupt. . . enemies': C.H. Carlisle, 'Serving My Time in Crystallography at Birkbeck.'

SIX *Woman of the Left Bank*

'French speciality': JC interview with VL, 14 Jun. 1985.
'It was a technique': Dr M. Oberlin, memorandum on Rosalind's experience at the Laboratoire Central des Services Chimiques de l'Etat; memo courtesy of A. Oberlin.
'At great risk to himself': author's interview with Rachel Glaeser, 4 Apr. 2000.
'knowing who was alive': author's interview with Marianne Weill Baruch, 11 Oct. 1999.

'unEnglished': D.H. Lawrence, *Mr Noon*, pp. 134–5.
'France's No. 1': D. Bair, *Simone de Beauvoir*, p. 327.
'the owner is': RF to parents, early 1947.
'while London mists': RF to parents, ibid.
'Of course my standard': RF to EF, 4 May 1947.
'pure research': ibid.
'One only feels rich': ibid.
'the newly nationalised': the basic holiday travel allowance for an adult was £100, Apr. 1946–47, falling briefly to £35 in autumn 1947, then to nil. (Source, Sanctions Unit, Bank of England Press Office.)
'liberal, Cartesian': Dr M. Oberlin to Prof. Michio Inagaki, 'Rosalind E. Franklin – Who was She?', published in *Energeia*, p. 4.
'sound and cheerful': ibid.
'time of her life': ibid.
'I've been told': RF to parents, 12 May 1948.
'I've had dress material': RF to parents, 11 Oct. 1947.
'post-war style': Jane Mulvagh to author, 7 Nov. 2001.
'delightfully clean': RF to parents, 25 Jul. 1947.
'Bananas suddenly appeared': RF to parents, 27 Jan. 1948.
'As for your remark': RF to EF, 4 May 1947.
'golden hands': author's interview with VL, 26 Oct. 1998.
'When Vittorio and Rosalind': AS interview with Dr June Goodfield, 1981, ASA.
'He was Jewish': author's interview with VL, 26 Oct. 1998.
'*Toutes les jeunes filles*', author's interview with Denise Tchoubar, 4 Apr. 2000.
'At the same time': AS interview with Mering, 28 May 1970, ASA, and author's interview with Rachel Glaeser, 4 Apr. 2000.
'puritanical': AS interview with Rachel Glaeser, 28 May 1970, ASA; Glaeser actually used the word 'protestant'.
'terrible, hearty': AS interview with Adrienne Weill, 11 Jun. 1970, ASA.
'Something *happened*': author's interview with UR, 20 Jan. 1999.
'some sort of sexual': author's interview with Anne Piper, 13 Jan. 1999.
'something': AS interview with Jacques Mering, 28 May 1970, ASA.
'We started out': RF to CF, 16 Aug. 1947, in Glynn, op. cit., p. 276.
'in black hat': RF to parents, 3 Apr. 1948.
'I've just had an absurd': RF to parents, 12 May 1948.
'Realities in Palestine': *Economist*, 27 Mar. 1948.
'seen the 1914 war': RF to EF, 26 Mar. 1948.
'the absence of': RF to parents, 9 Sep. 1947.
'to avert': Evelyn Waugh, *Vile Bodies*, p. 16.
'with a good five minutes': RF to parents, 'Wednesday', 1948.
'It's not dirt': Dr M. Oberlin, memo on Rosalind's experience at the Laboratoire Central des Services Chimiques de l'Etat.
'her radiation monitoring badge': G. Friedman and M. Friedland, *Medicine's 10 Greatest Discoveries*, p. 209.
'Requisitioned by the Germans': author's interview with Rachel Glaeser, 23 Mar. 2000.
'Nor, come to that': CF to author, 31 Jul. 2000.
'utterly exhausted': RF to parents, 'Hotel di Fango, Wednesday'.

'she settled her sleeping bag': Mering retold this story *en riant* to, among others, Denise Tchoubar: author's interview with Tchoubar, 4 Apr. 2000.

'was like Queen Victoria': Rachel Glaeser to AS, 28 May 1970, ASA.

'most unpleasant': RF to parents, 2 Sep. 1948.

'Nor, come to that': CF to author, 31 Jul. 2000.

'Denise observed': AS interviews with Denise Luzzati, 27 and 29 May 1970, ASA.

Agnes Oberlin, for her part, believed that Mering wanted to be loved and would have returned any love he felt directed to him. Rosalind, in her view, was one of three examples of Mering accepting the boundless admiration and slavish devotion of one of his female staff, only to reject them swiftly and with rancour at the first sign of independence or intellectual discord. Oberlin to BM, 12 May 2001.

'Rosalind burned': ibid.

'very crisp and very pretty': AS interview with Dr June Goodfield, April 1981.

'white-shirt, dark-skirt': fashion interpretation by Jane Mulvagh, 2 Nov. 2001.

'I wish you wouldn't': RF to parents, 'Thursday'.

'it would be a *bad* idea': RF to Muriel, 'Monday'.

'I can't face the man': RF to parents. 'Tuesday'.

'Don't you realise?': R. Olby in *Daedalus*, p. 244.

'entirely reasonable': Charles Coulson to RF, 11 Jun. 1949, CAC.

'Rosalind certainly planned': Margaret Nance Pierce to author, 22 Jan. 2001.

'also claimed descent': CF to author, 8 Oct. 2001.

'I shan't come home': RF to CF, 10 Oct. 1949.

SEVEN *Seine v. Strand*

'If you are interested': Coulson to RF, 11 Jun. 1949 in Olby, *The Path to the Double Helix,* p. 345.

'I am, of course': RF to Coulson, 23 Mar. 1950 in ibid., pp. 345–6.

'Philip, an American war veteran': author's interview with Philip Hemily, 3 Nov. 1998.

'Monsieur J. Mering, for his guidance': R. Franklin, 'The Interpretation of diffuse X-ray Diagrams of Carbon.'

'half of me': RF to parents, 3 Mar. 1950.

'The best I ever': RF to parents, 15 May 1950.

'Denise Luzzati could see': AS interviews with Denise Luzzati, 26 and 28 May 1970, ASA.

'Rosalind's sister and brother': Jenifer Glynn to author, 27 Nov. 2001.

Rosalind, in her view, A. Oberlin to author, 12 May 2001.

'by means of X-ray diffraction': I.C.M. Maxwell to Principal, King's College, 7 Jul. 1950, CAC.

'I'd really like': RF to parents, 15 May 1950.

'pretty well top of the list': JTR to Principal, King's College, 19 Jun. 1950, CAC.

'not pleased': author's interview with Noel Richley, 20 Jan. 1999.

'there is a much healthier': RF to parents, 1 Aug. 1950.

'oscillate wildly': RF to CF, 26 May 1950.

'to change, "Seine. . . Strand': RF to parents, 13 Jul. 1950.

'I have no idea': RF to parents, 21 Aug. 1950.

'Of course': JC interview with VL, 14 Jun. 1985.

'I can't possibly': RF to parents, 29 Oct. 1950.
'I spend half': ibid.
'What depressed me most': RF to Evi Ellis, 1 Oct. 1950.
'Please forgive me' and preceding information: RF to JTR, 24 Nov. 1950, CAC.
'After very careful consideration': JTR to RF, 4 Dec. 1950, CAC. The existence of this important letter was not known until it appeared in 1974 in Olby, op. cit., p. 346.
'Dr Stokes . . . This means': ibid.
'foreigners better': RF to parents, 29 Oct. 1950.

PART TWO

EIGHT *What Is Life?*

'Published with other measurements': W.T. Astbury, and Florence O. Bell, 'X-Ray Study of Thymonucleic Acid'.
'I have not published': Oswald Avery to Roy Avery, 26 May 1943 in J. Cairns, G. Stent and J.D. Watson, *Phage and the Origins of Molecular Biology*, pp. 185–7.
'nucleic acids must': O. Avery, C. MacLeod and M. McCarty, 'Studies on the Chemical Transformation of Pneumococcal Types', pp. 137–59.
'doing something': E. Schrödinger, *What is Life*? p. 70.
'When one of the inventors': G. Stent, 'That Was The Molecular Biology That Was', *Science*, 160, 26 Apr. 1968; reproduced in Cairns et al. op. cit., p. 347.

NINE *Joining the Circus*

'great personal quarrels': H.F. Judson, *The Eighth Day of Creation*, p. 101.
'patriarchy personified': S. Benstock, *Women of the Left Bank*, pp. 447–8.
'politically embittered': M. Franklin, 'Rosalind', p. 20.
'tense and unbending': author's interview with John Bradley, 28 Nov. 2000.
'every time an Englishman': Alan Jay Lerner and Frederick Loewe, 'Why Can't the English?': Act 1, Song 2, *My Fair Lady*, drawn from G.B. Shaw's Preface to *Pygmalion*.
'typically upper-class': AS interview with Jean Hanson, ASA.
'Professor of Dogmatic Theology': Mary Fraser to author, 16 May 2000.
'the joint': Douglas Johnson, obituary of Lord Annan, *Guardian*, 23 Feb. 2000.
'hooded crows': Dr Walter Gratzer, conversation with author.
'*Chère Mademoiselle*': postcard from Faculté des Sciences de l'Université de Paris, Laboratoire de Minéralogie (signature illegible), 2 Feb. 1951.
'the PhD slave boy handed over in chains': Sayre, op. cit., p. 101.
'JT' etc: author's interview with RG, 19 Oct. 1999.
'beautiful dark eyes': JC interview with RG, 19 Apr. 1985.
'to concern himself': JTR to RF, 4 Dec. 1950, in Olby, op. cit., p. 346.
'very attractive': author's interview with Louise Heller, 9 Jan. 2000.
'Gosling, working in conjunction': JTR to RF, 4 Dec. 1950.
'grab her and get her in on the DNA work': author's interview with MW, 4 Nov. 2000, and S. Chomet, *Genesis of a Discovery*, p. 15.
'a single DNA fibre': ibid., p. 17.
'equipment from Paris': Etablissements Beaudouin to RF, 16 Feb. 1951, JNC.
'communicated by': R. Franklin, 'Crystallite growth in graphitising and non-graphitising carbons', pp. 196–218.

'rough edges': M. Wilkins, 'John Turton Randall', *Biographical Memoirs*, pp. 510–11.
'most valuable cargo': ibid., p. 505.
'put in his claim': JTR correspondence with his lawyers, the Ministry of Defence and the Royal Commission on Rewards for Investors, CAC.
'Endorsement for the unit': the history of the King's Biophysics Unit is told in Olby, op. cit., p. 329.
'beef steak between the poles of his magnet': Wilkins, op. cit., pp. 510–11.
'Randall's Circus': this was occasionally a term of abuse, JC interview with RG, 19 Apr. 1985.
'love of plants': Wilkins, op. cit., p. 510.
'tough as old leather': JC interview with MW, 17 Jun. 1985.
'a dog's walking': J. Boswell, *Life of Johnson*, p. 114.
'only seven': O. Opfell, p. xiv; *The Lady Laureates* J. Mason, 'Women Fellows' Jubilee', p. 133.
'a quasi-religious': M. Wertheim, p. xiv.
'barriers to the entry': ibid., p. xv.
'Harvard. . . cover of darkness': ibid., p. 223.
'command performance': author's interview with Heller, 9 Jan. 2000.
'steady watchful dark eyes': Wilkins, *Memoirs*, pre-publication ms.
'He discussed with her a paper': ibid.
'spikiness', 'real cream': ibid.
'trust Avery': Chomet, op. cit., p. 13.
'judged too severely': J.T. Randall, 'An Experiment in Biophysics', p. 2.
'young Norwegian researcher': ibid., p. 16.
'necking': ibid., p.15.
'This breaking down': ibid., p. 2.
'severe cramps': University College Hospital record, AD1651, notes by Prof. Nixon, H.S., 2 Sep. 1956.
'Oh well': JC interview with Freda Ticehurst, 31 May 1985.
'Very well read. . . bra undone': JC interview with Dr Simon Altmann, 26 Jun. 1985; author's interview with Altmann, 2 Dec. 1999.
'not only took my lecture': AS to GCD, 28 Jun. 1978, ASA.
'Camembert': Margaret Nance Pierce to author, 24 Jul. 1999; 'potatoes': Liebe Klug, 7 Oct. 1999; 'garlic': author's interview with UR, 29 Jan. 1999.
'Is it your turn': author's interview with UR.

TEN *Such a Funny Lab*

'Thanne longen folk': G. Chaucer, *Canterbury Tales*, Prologue.
'foreign visits': *MRC Annual Report*, 1951–52, p. 41.
'happy to stand in': Wilkins, op. cit.
'fantastically irregular': Watson, *The Double Helix*, p. 23.
'if Maurice really': ibid., p. 24.
'a natural structural system for biological macromolecules': Carlisle, op. cit., p. 29.
'after having suffered': S. Furberg, 'My Work on Nucleic Acid Structure 1947–49'.
'super': author's interview with RG, 19 Oct. 1999.
'She shouldn't': JC interview with Geoffrey Brown, 2 May 1985.
'had the sort of': author's interview with Louise Heller.
'Monsieur J. Mering': R. Franklin, 'Structure of Graphitic Carbons', pp. 253–61.
'a rather peevish': R.F. Tuckett to AS, 8 Apr. 1976, ASA.
'He's so middle-class': VL to author, 26 Oct. 1998.
'at least what one wants': F. Jacob, *The Statue Within*, pp. 262–3.
'the biggest mistake': F.H.C. Crick, *What Mad Pursuit?*, p. 58.

'Helices were': ibid., p. 60.
'*the* solution': ibid.
'the people giving': Richard Morrison, 'What did the Festival of Britain organisers get so right?', *The Times*, 26 Apr. 2001.
'its scientific genius': ibid.
'Go back to your microscopes': author's interview with MW, 4 Apr. 2000; also JC interview with MW, 17 Jun. 1985; Watson, op. cit.
'as far as the experimental X-ray effort': JTR to REF, 4 Dec. 1950, CAC.
'Decades later': author's conversation with MW.
'exciting project in biophysics': Wilkins, op. cit.
'Now she's trying': JC interview with G. and A. Brown, 2 May 1985.
'Her manner was brusque': Norma Sutherland Clarke in e-mail to author, 12 Jul. 2000.
'Anne knew': A. Piper, 'Camping Holiday 1951', p. 7. privately circulated memoir; also mentioned in A. Piper, 'Light on a dark lady', p.152.
'a heavenly view': RF to AW, 21 Oct. 1951, ASA.
'one of those very able': Piper, op. cit., p. 151.
'Brazenly he wrote': LP to JTR, 25 Sep. 1951, PA.
'Wilkins and others': JTR to LP, 28 Aug. 1951, PA.
'worth perhaps 1/5000 of a Nobel Prize': Chomet, op. cit., p. 42.
'Crick's reply': Friedman and Friedland, op. cit., p. 210.
'Stokes has supplied': MW to RF, in Olby, op. cit., p. 342.
'most beautiful X-ray photographs': J.D. Bernal, 'Obituary, Rosalind Franklin', *Nature*, and M. Franklin, op. cit., p. 27. With hindsight, according to Aaron Klug, earlier photographs taken at King's had obtained fuzzy patterns that could later be recognised as the B form of DNA. Rosalind's unique achievement was twofold, 'to obtain a well-defined B pattern and characterise it as belonging to a definite structural state of DNA': A. Klug, 'Rosalind Franklin and the Discovery of the Structure of DNA', p. 880.
'How dare you interpret my data for me!': in Olby, op. cit., p. 344.
'weeping': C. Franklin, op. cit., p. 6.
'I'm sorry I have been silent': RF to AW, 21 Oct. 1951, ASA.
'Far from settling': author's conversation with MW, 3 May 2000.
'acting as Envoy': RG to JTR, 15 Aug. 1972, CAC.
'He got very down': author's interview with Freda Ticehurst Collier, 14 Dec. 1999.
'It's just like snot!': RG to JTR, 15 Aug. 1972, CAC.
'We ended up smelling': R.D.B. Fraser to author, 31 Mar. 2001.
'the borderline between': Crick, op. cit., p. 20.
'this side of eccentric': JC interview with G. and A. Brown.
'she wore white': author's interview with RG, 19 Oct. 1999.
'To Dorothy Hodgkin': G. Ferry, *Dorothy Hodgkin: A Life*, p. 246.
'That's the way we did it': R.D.B. Fraser to author, 10 May 2000.
'Rosy's Parlour': JC interview with W.R. Seeds, 4 Jun. 1985.
'little schoolboys': ibid.
'When Fraser asked her': R.D.B. Fraser to author, 6 Feb. 2000.
'Wilkins said the same': ibid.
'That's very nice': author's interview with RG, 19 Oct. 1999.
'Momentarily': Watson, op. cit., p. 45.
'Though her features': ibid.

'the product': ibid.
'Stimulated': Olby, op. cit., p. 357.
'pugnaciously assertive': Watson, op. cit., p. 60. Watson ascribes this reaction to Gosling too.
'now she was a little bit': RG to JC.
'in fact there was a lot of water': Olby, op. cit., p. 360.
'This pact': Friedman and Friedland, op. cit., p. 223, 'and Watson, *The Double Helix* pp. 13–14.
'This interpretation': Joan Mason to author, 20 Aug. 1999.
'Alex Stokes had an equation': this ditty, performed at the 1952 Biophysics dinner, was written by and supplied by Stan Bayley and Prof. Edward Deeley.
'Over the holiday break': RF to AS, 1952, ASA.

ELEVEN *The Undeclared Race*
'It's what a crystallographer': author's interview with VL, 26 Oct. 1998.
'Crystallographers, notably Dorothy Hodgkin': Dorothy Hodgkin note to AS, enclosed in letter to David Sayre, 6 Jan. 1975, ASA.
'Adrienne Weill was well aware': AS interview with AW, 11 Jun. 1970, ASA.
'Real test': RF notebook labelled 'Stockholm 1951', FRNK 3/1, CAC.
'the structure speaking': author's interview with L.H.L. Cohen, 2 Oct. 1999.
'the queen of the Patterson': Ferry, op. cit., p. 244.
'Gosling had nightmares': R. Gosling in Chomet, op. cit., p. 66.
'That is what she felt she was there to do': ibid., p. 48.
'a helical structure': ibid, p. 4.
'was also derived': W. Cochran, F.H.C. Crick and V. Vand, 'The Structure of Synthetic Polypeptides . . .', p. 582.
'What is it': RF to AS, 1 Mar. 1952, ASA.
'retreating into a disconcerting': Glynn, op. cit., p. 276.
'watchful, suspicious and timid': Sir W. Scott, op. cit., Chapter 11.
'stacks'. . . 'homicide': AS to RF, 8 Mar. 1952, ASA.
'Whatever one may have against': RF to AS, 1 Mar. 1952, ASA.
'I went and took it back': author's interview with Geoffrey Brown, 10 Feb. 2000.
'She nearly scared': author's interview with Sir John Cadogan.
'She looked beautiful': AS interview with RG, ASA.
'Franklin barks': MW to FHCC, 1952, in Olby, op. cit., p. 366.
'Her thick sheaf of notes': R. Franklin, '5 lectures on X-ray diffraction', Jan-Feb. 1952, JNC.
'Like most British labs': G. Ferry quoting Warren Weaver of the Rockefeller Foundation, op. cit., p. 246.
'best she had ever seen': ibid., p. 275.
'handedness': ibid., pp. 275–6.
'Watson has often used': author's interview with Jack Dunitz, 2 Oct. 2000.
'From the X-ray evidence': MRC 52/200, PRO 72037, Ref. FD1–7102. Randall's progress report was one of two items on the agenda.
'would not take no': Margaret Nance Pierce to author, 26 Jul. 2000.
'"Don't" was Rosalind's advice': author's interview with Pauline Cowan Harrison, 26 Jul. 2000.
'extremely excellent. . . try our luck': JDW to Max Delbrück, 20 May 1952, in Olby, op. cit., pp. 367–8.
'interest whetted': ibid., p. 378.
'the cylindrical Patterson function': Olby, 'Rosalind Elsie Franklin',

Dictionary of Scientific Biography, pp. 139–42.
'As Rosalind was necessarily involved': Dorothy Hodgkin to David Sayre, 7 Jan. 1975, ASA.
'I'm afraid we always': FHCC in Judson, op. cit., p. 141.
'a pathetic place': RF to AS and David Sayre, 2 Jun. 1952, ASA.
'I do not know': Katarina Kranjc to AS, 24 Mar. 1970, ASA.
'They all spoke French': Prof. Dr D. Grdenič to AS, 11 May 1970, ASA.
'They reckon': RF to AS and David Sayre, 2 Jun. 1952, ASA.
'all too good': ibid.
'The pure sodium salt': 'The New Physics and Engineering Laboratories', King's College London pamphlet issued for the opening ceremony by the Lord Cherwell, FRS, 27 Jun. 1952.
'the most beautiful buildings': Watson, op. cit., p. 28.
'the most attractive site': J.D. Watson, *A Passion for DNA*, p. 21.
'two pitchmen': E. Chargaff in Judson, op. cit., p. 144.
'had to go': Dr Lewis Wolpert in 'The Dark Lady', BBC Radio 4, 27 Feb. 1987, said that Rosalind was asked to leave King's.
'continue to study': I.C. Maxwell to JTR, 1 Jul. 1952, CAC.
'banana-shaped': RF notebooks, 'banana-shaped peaks'.
'buried in paper': JC interview with W.R. Seeds, 4 Jun. 1985.
'We spent *ages*. . . *tell* us the structure': author's interview with RG, 19 Oct. 1999.
'Rubbish!': author's interview with AK, 6 Jul. 1999, and KCH to JC, 23 Jun. 1985.
'By the time': RG in Chomet, op. cit., p. 68.
'It is with great regret': RF and RG, 'Death of a Helix', in Judson, op. cit., p. 144.
'At no time': RG to JTR, 15 Aug. 1972, CAC.
'Stokes later admitted': JC interview with A. Stokes, 21 Jun. 1985.
'Why not ask': H.R. Wilson, 'The double helix and all that', p. 276.
'Colleagues at King's': JC interview with W.R. Seeds, 4 Jun. 1985; JC interview with RG, 24 Jun. 1985.
'She was a very powerful personality': author's interview with Pauline Cowan, 26 Jul. 2000.
'woollen underwear': LP to PP, 22 Oct. 1952, PA.
'Mama wants to know': LP to PP, 22 Oct. 1952, PA.
'full of French girls': PP to LP, 28 Oct. 1952, PA.
'Stranded whales': PP to L and AHP, 13 Jan. 1953, PA.
'with certainty': RF, in MRC report, 5 Dec. 1952, p. 8, PRO FD1/7102 66670.
'lunch . . . tea': PRO FD1/7102, 29 Dec. 1952.
'Salisbury is now satisfied': memo, Sir Harold Himsworth, 20 Dec. 1952 PRO FD1/7102 66670; not marked 'Confidential', Sir H.P. Himsworth to MP, 12 Jul. 1968, JNC.

TWELVE *Eureka and Goodbye*

'Rosy, of course': J.D. Watson, *The Double Helix*, p. 105.
'splendid X-ray photographs': R. Corey to RF, 13 Apr. 1953, in Olby, op. cit., p. 396.
'The stage reached': A. Klug, 'Rosalind Franklin and the Discovery . . .', pp. 808–10, 843–4; reprinted in Watson, op. cit., p. 156.
'You know how children': PP to LP, 13 Jan. 1953, PA.
'Second sunny day': PP to LP, early 1953, PA.

'I guess': 'Why Pauling didn't solve the structure of DNA', Jim Lake, letter to *Nature*, 1 Feb. 2001, Vol. 409.

'the A form of the molecule was not helical': Wilson, *TIBS* 13, 7, p. 277.

'Here's the Dean': Author's interview with Geoffrey and Angela Brown, 10 Feb. 2000.

'Since the door was already': Watson, op. cit., p. 95.

'not in the least': Carlisle, op. cit., p. 40.

'I was more aware': Watson, op. cit., p. 96.

'She nearly terrified': author's interview with Sir John Cadogan.

'She complained to a friend': author's interview with Dr Simon Altmann, 19 Jan. 1999; letters to author 23 Jun. 2000 and 4 and 5 Mar. 2001. Because Altmann returned to Argentina between early 1952 and the spring of 1953, he cannot place the date when she told him of her suspicions that someone had gone through her desk.

'they should have protected her': ibid.

'three papers': LP to PP, 10 Mar. 1953, PA.

'relatively free': R. Franklin and R. Gosling, 'Molecular Configuration in Sodium Thymonucleate', pp. 740–1; Watson, op. cit., p. 254.

'she had made a similar lunge': J.D. Watson, ibid., p. 96.

'She's got a very good B': author's interview with MW.

'Maurice had a perfect right': author's interview with RG; also F.H.C. Crick letter to author, 12 Apr. 2000: 'It seems to me Maurice did nothing wrong in showing the photo to Jim.'

'The instant I saw': Watson, op. cit., p. 98.

'Francis would have to agree': ibid., p. 99.

'According to Crick': Judson, op. cit., p. 167.

'We missed out': author's interview with H.R. Wilson, 17 Aug. 2000.

'Why not': F.H.C. Crick, *What Mad Pursuit?*, p. 70.

'I will tell you all': Olby, op. cit., p. 401, based on MW to FHCC 1953.

'I was inexperienced': M. Perutz, letter to *Science*, 27 Jun. 1969, pp. 1537–58, reprinted in Watson, op. cit., p. 209. Perutz's large archive of gathered evidence that the report was not confidential is in the JNC. It shows how stung Perutz, a highly ethical man, was by the charge.

'RF Notes for Feb. 23 1953': CAC.

'The nucleic acids': L. Pauling and R. Corey, 'A Proposed Structure for the Nucleic Acids', pp. 84–97.

'tight squeeze': LP to PP, 18 or 19 Feb. 1953, PA.

'enol was *out*': Jerry Donohue to Judson, 5 Mar. 1976, HFJA.

'Nearly home': A. Klug in Judson, op. cit., pp. 172–3.

'Base interchangeability': A. Klug, 'Rosalind Franklin and the Discovery of the Structure of DNA', pp. 808–10, 843–4, reprinted in Watson, op. cit., p. 157.

'It is easy to feel': Judson, op. cit., p. 172.

'Francis winged': Watson, op. cit., p. 115.

'Let's face it': J. Donohue to Judson, 5 Mar. 1976, HFJA.

'Morris Wilkins': PP to LP, 14 Mar. 1953, PA.

'I think you will be interested': MW to FHCC, 'Saturday', 7 Mar. 1953, in Olby, op. cit., p. 414.

'I felt the model *as such* was their work': Callander, p. 16.

'I told *everybody*': author's conversation with MW.

'I'm not wanted': JC interview with Freda Ticehurst Collier, 31 May 1985.

THIRTEEN *Escaping Notice*

'The model was so beautiful': author's conversation with JDW, 5 Jan. 2002.
'of not working': Olby, op. cit., p. 368.
'I think you're a couple': MW to FHCC, 18 Mar. 1953, in Olby, op. cit., pp. 417–18.
'being a trifle awkward': ibid.
'we rather stopped him': ibid.
'like a scalded rat': JC interview with W.R Seeds, 4 Jun. 1985.
'quite upset': JC interview with RG, 24 Jun. 1985.
'everyone': JC interview with Anthony and Margaret North, 11 Apr. 1985.
'Had it': Jerry Donohue to AS, 19 Dec. 1975, ASA.
'Get writing!': author's interview with RG.
'cabled': Bruce Fraser to author, 10 Feb. 2000.
'in the press': J.D. Watson and F.H.C. Crick, 'A Structure for Deoxyribonucleic Acid', pp. 737–8; Watson, op. cit., pp. 237–41. Crick thought Fraser's model 'feeble', FHCC to author, 9 Jan. 2002.
'the following communications': Watson and Crick, op. cit.
'obvious': FHCC to author, 16 Dec. 2001; HFJ on MP to author, 16 Dec. 2001, and RG to author, 30 Dec. 2001.
'Delete *very beautiful*': MW to FHCC, 18 Mar. 1953, in Olby, op. cit., p. 418.
'Thus our general ideas': R. Franklin and R. Gosling, 'Molecular Configuration in Sodium Thymonucleate', pp. 740–1; Watson, op. cit., p. 256.
'It has not escaped': J.D. Watson and F.H.C. Crick, 'A Structure for Deoxyribonucleic Acid', p. 737.
'Where's Rosalind?': author's interview with Freda Ticehurst Collier, 14 Dec. 1999.
'You will no doubt': JTR to RF, 17 Apr. 1953, CAC.
'Science differs': W. Gratzer, *The Oxford Book of Scientific Anecdote*.

FOURTEEN *The Acid Next Door*

'to get her assistance': JDB to A. Tattersall, 1953, FRNK 2/31, CAC.
'in the principles of the arts': R. Furth, 'The Physics Department of Birkbeck College', p. 150.
'discourses on the Cretan Renaissance': C.P. Snow, *The Search*, Part III, Chapter II, pp. 180–1.
'eulogised Stalin': B. Swann and F. Aprahamian, *J.D. Bernal*, p. 151.
'It took over the': Carlisle, op. cit., p. 20; also Furth, op. cit., p. 153.
'becoming unbalanced': J.F. Lockwood to JDB, Sep. 7 1954.
'Who Have Not': author's interview with Dr Wolfie Traub, 20 Jul. 1999. I am indebted to Traub for this anecdote, confirmed in an e-mail to author, 26 Jul. 1999.
'one of his recruits': author's interview with Dr John Mason, 22 Oct. 2001.
'to get her assistance': JDB to A. Tattersall, 1953, FRNK 2/31, CAC.
'Randall's ban': JTR to RF, 17 Apr. 1953, FRNK 13, CAC.
'I am deeply': R. Gosling, 'Abstract of Thesis submitted by R.G. Gosling for the PhD Examination'.
'the perfect front': Watson, op. cit., p. 67.
'because if': FHCC to RF, 5 Jun. 1953, CAC.
'over my head': Jacob, op. cit., p. 265.
'baloney as well': G. Stent to author, 27 Jan. 2000.

'un-twiddle-ase': JC interview with RG, 24 Jun. 1985.
'I'm Watson': this ditty, 'Up the Krick with Watson' by E.S. Anderson, curiously spells Crick as 'Krick', a nicety inaudible to the audience at the Cold Spring Harbor Symposium in 1953, HFJA.
'And they're neutered': author's interview with Dr June Goodfield, 29 Feb. 2001.
'He did not know what "helical" meant': S. Furberg to Judson, 16 Mar. 1976, HFJA.
'an almost simultaneous': J.D. Bernal, 'The Material Theory of Life', p. 325.
'She makes my clock work': Piper, op. cit., p. 153.
'I find it hard': RF to MF, 8 Aug. 1953.
'be tempted': RF to MF, 21 Aug. 1953.
'I can sympathise': RF to parents, 1 Sep. 1953.
'I think it shook': Irene Neuner to Anne Sayre, 22 Jan. 1974, ASA.
'The only "restaurant"': RF to parents, 1953.
'I had been more than adequately': RF to MF, 21 Aug. 1953.
'Ros wouldn't know': author's interview with UR, 20 Jan. 1999.
'why had she never married': Neuner to Anne Sayre, 22 Jan. 1974, ASA.
'wonderful': RF to parents, 9 Sep. 1953.
'one of the cameras': JDB to JTR, 18 Oct. 1953, CAC.
'We do in fact': JTR to RF, 4 Nov. 1953, CAC.
'the tobacco mosaic virus': See *Phil. Trans. Royal Society*, Mar. 1999, p. 521. The entire issue is devoted to the tobacco mosaic virus.
'getting a virus': information from P. Butler and A. Klug, 'The Assembly of a Virus'.
'the way to DNA': Watson, op. cit., p. 75.
'For myself': RF to AS and David Sayre, 17 Dec. 1953, ASA.
'Rosy's Parlour': the programme did not spare Randall either, 'If you have enjoyed your dinner, spare a copper for St Paul, Friends of St Paul's, The Deanery, St Paul's, EC4.'

FIFTEEN *O My America*

'The Gordon organisers': FHCC to RF, 29 Apr. 1954, FRNK, 2/33, CAC.
'Since leaving': RF to LP, 1954, PA.
'similar work': RF to Prof. Pollard. 13 Apr. 1954, CAC.
'There is considerable objection': A.C. Fieldner to J.W. Garland, 24 Feb. 1954, CAC.
'One would see': Martyn Pease to author, 6 Mar. 2000.
'Julian Alps': Dusan Hazi to author, 2 Nov. 1999.
'Stopped one hour': RF to MF, 'Prestwick, Saturday 2 a.m.', 1954.
'of America as having trees': RF to CF and Charlotte Franklin, 1954.
'actually a University': RF to MF, 31 Aug. 1954.
'if she should tip': author's interview with Alex and Jane Rich, 17 Apr. 1999.
'a family of summer residents': Barbara Little to author, 24 May 2000.
'the bemused look of an English poet': J.D. Watson, *Genes, Girls and Gamow*, p. 88.
'that she had been judged': ibid., pp. 94–5.
'which I was beginning': RF to CF and Charlotte Franklin, 31 Aug. 1954.
'uncivilised and rude': RF to MF, 14 Sep. 1954.
'I'm also astonished': RF to CF and Charlotte Franklin, 31 Aug. 1954.
'Here, apart from hurricanes': ibid.

'My lab visits': RF to MF, 15 Sep. 1954.
'Pittsburg is as black': RF to MF, 'State College', 14 Sep. 1954.
'a comedy': RF to MF 24 Sep. 1954.
'The front door': RF to MF, Wisconsin, postmark 23 Sep. Chicago, 1954.
'In their museums': RF to MF, 14 Sep. 1954.
'the Indians as a tourist attraction': RF to parents, Sep. 1954.
'The oldest influence': RF to MF, 3 Oct. 1954.
'about 25 miles': RF to parents, re: Pasadena.
'A less successful outing': Sydney Brenner to author, 24 Jan. 2000.
'the first unfriendly': RF to parents.
'Watson and I': F.H.C. Crick, 'The Structure of the Hereditary Material'. Another very subdued acknowledgement to her crucial part in the great discovery appeared in the Watson-Crick article in mid-1954 in the *Proceedings of the Royal Academy*.
'the pleasantest possible memories': RF to LP, 19 Oct. 1954, PA.

SIXTEEN *New Friends, New Enemies*

'She needed a collaborator': Judson, op. cit., p. 172.
'old cronies': Jacob, op. cit., p. 262.
'No one could match': Watson, *The Double Helix*, p. 72.
'As you probably expected': N.W. Pirie to RF, 6 Dec. 1954, FRNK 2/33, CAC.
'Many thanks': RF to N.W. Pirie, 7 Dec. 1954, FRNK 2/33, CAC.
'There not being an ultracentrifuge': RF to Barry Commoner, 4 Mar. 1955, FRNK 2/33, CAC.
'Facts are facts': author's conversation with DC, 6 Dec. 1999.
'registered for a PhD degree': Carlisle, op. cit., p. 36; RF to KCH, 6 Jun. 1955, JNC.
'It takes imagination': AS interview with AK, and speech at St Paul's: 'She worked beautifully': Heather Brigstocke, 'Report by the High Mistress at Prizegiving', 27 Sep. 1983, *Paulina*, 1983.
'would have gone': JC interview with KCH, 23 Jun. 1985.
'We want to make': author's interviews with KCH, 24 Jan. 2000 and 30 Oct. 2001.
'C'est un endroit': VL interview with author, 26 Oct. 1998.
'She questioned his intellectual abilities': author's interview with Bryon Wilson, 8 Feb. 1999.
'We never looked back': author's interview with Stan Lenton, 13 Jul. 1999.
'if he might work with her': DC to RF, 9 Apr. 1955, FRNK 2/33, CAC.
'if that doesn't put you off': RF to DC, 19 May 1955, CAC.
'and she turned out': DC to author, 6 Dec. 1999.
'Hey, Ros!': KCH to author, email, 5 Aug. 2001.
'She didn't seem to know': Evi Wolgemuth to author, 11 Jan. 1999.
'She was a good aunt': author's interview with AK, 9 May 2000.
'girl of eighteen': W.G. Alexander to JDB, 2 May 1956, ARC archive/Cox.
'a rather concentrated solution': RF to Dr P. Kaesberg, 18 Jul. 1955, FRNK 2/33, CAC.
'Crick suggested': FHCC to RF, 3 Jun. 1955, FRNK 2/33, CAC.
'I've had a long talk': JDW to RF, 22 Jul. 1955, FRNK 2/33, CAC.
'cut back': A.J. Caraffi to RF, 15 Dec., 1955, FRNK, CAC.
'My age is 35': RF to JDB, 25 Jul. 1955, Birkbeck College Crystallography Laboratory, CAC.

'her annual salary': A.J. Caraffi to RF, 15 Dec. 1955, CAC.
'Slater refused': see RF's 'Notes on meeting with Slater', 29 Sep. 1955, CAC.
'exceptionally distinguished': Glynn, op. cit., p. 267.
'Presumably somebody': RF's notes on meeting with Sir William Slater, 29 Sep. 1955, CAC.
'we must remain dependent': R. Franklin, 'Progress Report of the Agricultural Research Council group in Birkbeck Crystallography Laboratory, for the year 1955 (accompanying application for renewal of grant)', FRNK 2/36, CAC.
'his position is such': ibid.
'Meeting Rosalind': author's interview with Dan Jacobson, 16 Aug. 2000.
'tea party': Prof. Dr D. Grdenič to AS, 11 May 1970.
'even capable': Fred Dainton to AS, 8 Nov. 1976, ASA.
'The wives': The Ciba Foundation programme for symposium on 'The Biophysics and Biochemistry of Viruses', 26–28 Mar. 1956, Novartis Foundation.
'As I have never been': RF to AW, 25 Feb. 1956, ASA.
'But, did anyone': author's interview with FHCC and Odile Crick, 22 Apr. 1999.
'is concerned with': R. Franklin, 'Note on the Future of the ARC Research Group in Birkbeck College Crystallography Laboratory', 9 Mar. 1956, CAC.

SEVENTEEN *Postponed Departure*
'Newcastle supplier': RF to R.H. Joyce, 29 May 1956, JNC.
'Her work': JDB to Sir William Slater, 18 Apr. 1956, FRNK 2/31, CAC.
'a table with a copious': RF to MF, 16 Jun. 1956.
'an abnormal patch': RF to MF, 24 Jun. 1956.
'but on an American scale': ibid.
'by quite a large amount': RF to MF, 24 Jun. 1956.
'However, as it's better': RF to AK, 21 Jun. 1956, JNC.
'Don is injecting': RF to AK, 25 Jun. 1956, JNC.
'His young staff': WLB to RF, 26 Jun. 1956, JNC.
'The quality and quantity': RF to MF, 15 Jul. 1956.
'Eh wot?': author's conversation with William Ginoza, 5 Mar. 2001.
'The answer': Betty Siegel to author, 9 Aug. 2001.
'a very brilliant ex-Italian': RF to parents, 15 Jul. 1956. This and subsequent details of the mountain trip from the same letter.
'sharp pains': AS to GCD, 23 Jun. 1976; also author's interview with Mair Livingstone, 15 Feb. 1999.
'or alternatively on bribing': RF to parents, 26 Jul. 1956.
'As on my last trip': ibid.
'Rosalind used to glow': Ethel Tessman to AS, 1 Jan. 1976, ASA.
'I'm going to see Don': RF to AK, 5 Aug. 1956, JNC.
'friend from the East': RF to MF, 9 Aug. 1956.
'Then, in spite of everything': ibid.
'chaste': author's interview with DC, 19 Dec. 1999.
'might have loved': Sayre, op. cit., p. 184.
'a delightful man': AS, two-page letter to John Simmons, archivist of All Souls College, Oxford; post-publication correspondence, ASA.
'Most telling for me': Irwin Tessman to author, 19 Dec. 1999.
'I've been waiting twenty years': author's interview with Dr June Goodfield, 27 Feb. 2001.

'*very* good looking': Caroline Carlson to author, 4 May 2000.
'You're not pregnant?': author's interview with Dr Mair Livingstone, 15 Feb. 1999.
'there is no reason': Dr Linken to Mr Norman Morris, 30 Aug. 1956, UCH.
'URGENT': W.C.W. Nixon, UCH Case No. AD 1651.
'treatment with ribonuclease': RF to Wendell Stanley, 30 Aug. 1956, JNC.
'a large proportion': RF to Dr Pomerat of Rockefeller Foundation, 31 Aug. 1956, JNC.
'the findings are most unfortunate': Prof. W.C.W. Nixon to Dr Linken, 5 Sep. 1956, UCH Case No. AD 1651.
'size of a croquet ball': Prof. Nixon, UCH notes for Miss Rosalind Franklyn [sic], 'Right oophorectomy and left ovarian cystectomy', Case No. AD 1651, 4 Sep. 1956.

EIGHTEEN *Private Health, Public Health*

'second operation': Prof. Nixon to Dr Linken, 3 Oct. 1956; also Histology report, 3 Oct. 1956, UCH Case No. AD 1651.
'agitated': AS notes following Jacques Mering interview, 28 May 1970, ASA.
'much in love with Mering': AS to Einar Flint, May/Jun. 1970, ASA.
'a truly immense': ibid.
'Use my flat': Kerlogue, op. cit.
'everything is going very well': RF to AS, postmarked 25 Oct. 1956, ASA.
'I think it is my bedtime': M. Franklin, op. cit., p. 20.
'female' and 'I'm afraid': Boston University Prof. I. Dorothy Raacke, Biological Science Center, to AS, 22 Jan. 1976, ASA.
'Reactions': RF to AS, 18 Mar. 1957, ASA.
'emotional and confused': RF to AS, 8 Oct. 1957, ASA.
'I have heard': JDW to AK, 13 Nov. 1956, JNC.
'Please give Dr Klug': MW to Biophysics Research Unit, 22 Nov. 1956, JNC.
'Recent work has shown': R. Franklin, 'Application for Research Grant E-1772, Department of Health, Education and Welfare', p. 3, JNC.
'The thing that impressed': RF to AS, 18 Mar. 1957, ASA.
'Dear Sage': Lord Rothschild to JDB, 25 Mar. 1957, ARC 253/57 Council Minutes, 19 Mar. 1957.
'I feel in the long run': FHCC to AK, 14 Dec. 1956, JNC.
'Rosalind is well': AK to Dr P. Newmark, University of Kansas, 27 Feb. 1957, JNC.
'the final year': W.C. Alexander to JDB, 17 Apr. 1957, ARC 177/57.
'rat liver nucleoprotein': Mary L. Petermann to RF, 28 Nov. 1956, JNC.
'The models, he said': R.W. Moulder to RF, 7 Oct. 1957, JNC.

NINETEEN *Clarity and Perfection*

'cobalt therapy': begun 15 May 1957, ended 14 Jun. 1957, radiotherapy record, UCH Case No. AD 1651.
'Observations on': Robley Williams, Friday evening discourse, Royal Institution, 14 Jun. 1957.
'There is absolutely': FHCC to RF, 23 May 1957, ASA.
'One so easily': MP to RF, postscript to previous letter.
'preliminarily': J. Palmer Saunders to RF, 9 Jul. 1957, JNC.
'pelvic mass': second opinion, entry for 4 Jul. 1957, radiotherapy record and medical history, UCH Case No. AD 1651.

'did not have the strength': author's interview with DC.
'glorious weekend in Zermatt': RF to AS, 8 Oct. 1957, ASA.
'This was my first': ibid.
'Luzzati's mother': VL to author, 20 Sep. 2001.
'In view of the extremely small': JDB to J.F. Lockwood, 4 Jul. 1957.
'You'll never guess': M. Franklin, op. cit., p. 17; Glynn, op. cit., p. 281.
'I am myself quite sure': Prof. E.T.C. Spooner to JDB, 10 Oct. 1957, ARC.
'From Oct. 57' RF curriculum vitae, FRNK, CAC.
'her stomach had swelled': author's interview with A. Piper, 13 Jan. 1999.
'chemotherapy': the Royal Marsden Hospital archives are not available.
'She had seen a parcel': GCD to AS, 9 Jun. 1976, ASA.
'Why are you going': author's interview with Nina Franklin, 12 Apr. 1999.
'my old Nurse': will of Rosalind E. Franklin, Central Probate Registry.
'What does Aaron need?': author's interview with Dan Jacobson.
'not *that* ill': JG to author, 20 Apr. 2001.
'*parasseuse*': RF to VL, 12 Feb. 1958, JNC.
'If you feel you want': WLB to MP, 12 Apr. 1958, JNC.
'waiting for some neutral glass': AK to Dr Taverne, 22 Apr. JNC.
'Still fighting': AK and RF to JDB, 18 Feb. 1958, Cruickshank archive.
'lack of heavy atoms derivating': AK to DC, 8 May 1958, JNC.
'most of the winter . . . over together': RF to DC, 16 Mar. 1958, JNC.
'Perutz came personally': interview with I. Hargittai, *Chemical Intelligencer*, Oct. 2000, p. 29.
'crawl up': AS to GCD, 23 Jun. 1976, ASA.
'five pages': RF scientific papers, notes for 28 Mar. 1958, JNC.
'when both she and I knew': GCD to AS, 31 May 1977, ASA.
'There is no doubt': ibid.
'with defiance': AS interview with Jacques Mering, 28 May 1970, ASA.
'He suspected': ibid.
'There she is': GCD to AS, 9 Jun. 1976.
'a brief flicker': [initial illegible] Haddow to JDB, 21 Apr. 1958, FRNK 2/31, CAC.
'She imagined': M. Franklin, op. cit., p. 15.
'I'm glad you've come': GCD to AS, 9 Jun. 1976, ASA.
Special Correspondent, 'Atomic Crystal Gazing in Brussels, Glamour of British Tradition', *The Times*, 15 Apr. 1958, p. 9.
'A Research Scientist': Rosalind E. Franklin, death certificate.
'His senses': W.B. Yeats, 'Ego Dominus Tuus', *Collected Poems*, p. 367.
'Rosalind Franklin, Virus Researcher': *New York Times*, 20 Apr. 1958.
'Rosalind Franklin's early and tragic death': J.D. Bernal, 'Obituary: Dr Rosalind Franklin', *The Times*, 19 Apr. 1958.
'As a scientist': J.D. Bernal, 'Obituary: Rosalind Franklin', *Nature*.
'She was very modest': Alice Franklin to AS, 24 Apr. 1958, ASA.
'Rosalind was always good': author's interview with AK, 6 Jul. 1999.

EPILOGUE *Life After Death*

'Concerning Rosalind': MW to JDW, 25 Jul. 1966, JNC.
'dowdier': Watson's phrase in *The Double Helix* is 'at the age of thirty-one her dresses showed the imagination of English bluestocking adolescent'.

'required the written consent': JDW to MP, 27 Sep. 1966, JNC.
'Francis Crick and Wilkins': JDW to MP, 26 Sep. 1966, JNC.
'In a fierce letter': LP to FHCC, 25 Apr. 1967, JNC.
'a rather hysterical reaction': Sayre, op. cit., p. 219.
'I was furious': M. Perutz, 'How the secret of life itself was discovered'.
'unfair to me': MW to T.J. Wilson, 4 May 1967, JNC. Wilkins's handwritten notes on an early version of Watson's *The Double Helix* shows that he tried to amend Watson's disparagement of her attractiveness, writing in the margin, 'often held that she was a very handsome girl', King's College London Archive. In retrospect, Wilkins acknowledges that there was more depth to Watson's book than he recognised at the time, and that the personal interactions of scientists merit recording.
'Harvard's Board': Watson, op. cit., p. xxii.
'I would rather': Dr June Goodfield, film proposal, 'Rosalind', quotes AK hearing this from MF.
'Professor Watson's Memoirs', *Nature*, 217, 23 Mar. 1968.
'communicates the spirit': G. Stent, 'A Review of the Reviews', in Watson, op. cit., p. 167.
'a decisive breakthrough': J.D. Bernal, 'The Material Theory of Life', p. 325.
'lines such as': J.D. Watson, *The Double Helix*, pp. 14, 15, 45.
'the scientist as human being': Mary Ellmann, in *The Double Helix*, from *Yale Review*, 57 (Summer 1968), pp. 631–5.
'Elizabeth Janeway attacked': E. Janeway, *Man's World, Woman's Place*, p. 111.
'had been exacerbated': C. Pert, *Molecules of Emotion*, p. 111.
'Base Pairs': Watson says he meant this interim title to be ironic, a riposte to Crick for disliking 'Honest Jim', JDW to author, Jan. 2002.
'quite horribly well done': 'Lucky Jim', *New Statesman*, 29 Nov. 1999, reprint of 1954 review, p. 62.
'Let's just start': JDW speech at Harvard, 30 Sep. 1999.
'never telling her': Crick says he doesn't recall ever actually telling Rosalind how crucial her work was to their discovery because he thought it was obvious (FHCC to author, 17 Dec. 2001). JDW in conversation with author acknowledged that with hindsight not telling her was a mistake, but once they had made the discovery, the model proclaimed its own meaning and it was clear that they hadn't needed her data. Working out the function of the base pairs was the essential step.
'public platforms': see below: Watson in person and Crick on video. At the dedication of the Franklin-Wilkins Building, King's College London, 22 Mar. 2000, Crick's words were, 'We could never have proposed the model but for what they [Wilkins and Franklin] told us of their results.'
'the real tragedy': David Harker, in R. Hubbard, 'The Story of DNA', Marjorie Senechal (ed.), *Structures of Matter and Patterns in Science*. Also, author's interview with David Sayre, 1 Oct. 1999. Also, Robert Sinsheimer, 'The Double Helix', review in *Science and Engineering*, Sep. 1968, p. 6, reprinted in *The Double Helix*, pp. 191–4.
'Penniless Heinz': Stent, op. cit.,

p. 161. See also Sayre, op. cit., p. 194: 'Was this why "Rosy" was invented? To rationalise, justify, excuse, and even to "sell" that which was done that ought not really to have been done?'

'Well dear': C. Franklin, op. cit.

'Jim's novel': author's interview with MW, 22 Mar. 1999.

'quite witty': author's conversation with RG.

'Rosalind's final, brilliant work': M. M'Ewen to HFJ, 15 Sep. 1976.

'Rosalind's difficulties': F.H.C. Crick, 'How to Live with a Golden Helix', in *The Sciences*, Sep. 1979, p. 7.

'The major opposition': ibid.

'analytical mind': AS interview with FHCC, 16 Jun. 1970. ASA.

'today would bring down the wrath': author's interview with KCH, 24 Jan. 2000.

'Ashkenazi gene': see for example U. Beller et al., 'High frequency of BRCA1 and BRCA2 mutations in Ashkenazi Jewish ovarian cancer patients, regardless of family history', *Gynecol. Oncol*, 67, 126–6 (1997); also E. Levy-Lahad et al., 'Founder BRCA1 and BRCA2 mutations in Ashkenazi Jews in Israel, frequency and differential penetrance in ovarian cancer and in breast–ovarian cancer families', *American Journal of Human Genetics*, 60, 1059–67 (1997).

'This entailed': Bryan Sykes, *The Seven Daughters*, p. 26.

'memorial fund': AK to DC, 2 May 1958; DC to AK, 5 May 1958, JNC. Also JG to author.

'So you were in love': RG to author, 19 Oct. 1999.

'The circumstances of her death': KCH to AS, 22 Apr. 1971, ASA. He expressed the same feelings in interviews with author, 24 Jan. 2000 and 26 Nov. 2001.

'nun-like': AK lecture on RF at RS, 18 Mar. 1999.

'droopy drudge': AS to Christopher Salazar, 16 Feb. 1993, ASA.

'is of a brilliant': Jon Bate and Hilary Gaskin, 'Unsung Pioneer', *New Statesman*, 8 Jul. 1983.

'Franklin never received': 'Fame at Last', in *The Times*, 10 Feb. 1992.

'In the spirit of righting old wrongs': in Mar. 2000 Wilkins wrote to Bruce Fraser in Australia and apologised for having failed to insist in 1953 that Fraser's note that he, Wilkins, had solicited on 17 Mar. 1953, his three-chain model of DNA, was not published in 1953 along with the Watson–Crick–Wilkins, Wilson, Franklin–Gosling papers in *Nature*.

'*Waarom kreeg*': 'Why did Rosy not get the Nobel Prize?', *Vrij Nederland*, 15 Aug. 1998.

'great joy' and subsequent quotes: WLB to MP, 22 Nov. 1962, JNC.

'Bragg wouldn't have': WLB to Erik Hulthen, chairman, the Nobel committee for Physics, 9 Jan. 1960, JNC.

'embarrassed': Ferry, op. cit., p. 289, citing M. Perutz, 'Forty years of friendship with Dorothy', in G. Dodson, J. Glusker and D. Sayre, *Crystal Structure Analysis: A Primer*, OUP.

'I have always felt': JTR to RG, 29 Aug. 1972, CAC.

'Had her life': A. Klug, *Les Prix Nobel en 1982*: Stockholm, pp. 94–5.

'She would have solved it': A. Klug, 'Rosalind Franklin and the Discovery of the Structure of DNA'; reprinted in Watson, *The Double Helix*, p. 154.

'debt of honour': In 2000, A. Klug also enjoyed considerable financial gain from the sale of his papers, which included some of Rosalind's notebooks, papers and letters, to the JNC.

'the world's most coveted intellectual award': see 'Kofi Annan, Nobel Laureate', letter to the editor of the *New York Times* by Ware G. Kuschner, assistant professor of medicine, Stanford University, 16 Oct. 2001.

'Avery should have gotten': author's interview with E. Chargaff, 18 Apr. 1999.

'collective enterprise': John Maddox, 'Science has changed and so must the Nobel Prize', *Independent*, 11 Oct. 2000

BIBLIOGRAPHY

Agar, Delia M., 'Some Comments on Rosalind Franklin and DNA', Newnham College Library No. 828.77, 30 Mar. 1979.

Anon, 'Dr Rosalind Franklin', *The Times*, 19 Apr. 1958, p. 3.

Anon, 'Twenty-one years of the double helix', *Nature*, 248, p. 721, 26 Apr. 1974.

Astbury, W.T., and Bell, Florence O., 'X-ray Studies of Nucleic Acids in tissues', *Symposia of the Society for Experimental Biology*, I, pp. 66–76 (Cambridge: Cambridge University Press, 1947).

——'X-Ray Study of Thymonucleic Acid', letter to *Nature*, 141, No. 3573, pp. 747–8, 23 Apr. 1938.

Avery, Oswald T., MacLeod, Colin M., and McCarty, Maclyn, 'Studies on the Chemical Transformation of Pneumococcal Types', *Journal of Experimental Medicine*, Vol. 79, pp. 137–58 (received for publication, 1 Nov. 1943).

Bailes, Howard, *Once a Paulina* (London: James and James, 2000).

Bair, Deirdre, *Simone de Beauvoir* (London: Jonathan Cape, 1990).

Bangham, D.H., and Franklin, Rosalind E., 'Thermal Expansion of Coals and Carbonised Coals', *Transactions of the Faraday Society*, 42B, pp. 289–94, 1946.

Bangham, D.H., Franklin, R.E., Hirst, W., and Maggs, P., 'A structural model for coal substance', British Coal Utilisation Research Association, *Fuel*, xxviii-10, pp. 231–7.

Baon, G.E., and Franklin, Rosalind E., 'The a[alpha] dimension of graphite', *Acta Cryst.*, 4, p. 561, 1951.

Beard, Mary, 'Bench Space', *London Review of Books*, pp. 27, 29, 15 Apr. 1999.

Benstock, Shari, *Women of the Left Bank: Paris, 1900–1940* (London: Virago, 1987; Austin: University of Texas Press, 1986).

Bentwich, Helen Franklin, *If I Forget Thee: some chapters of autobiography, 1912–1920* (London: Elek, 1973).

Bentwich, Norman, *The Jews In Our Time: The development of Jewish life in the modern world* (London: Penguin, 1960).

Bermant, Chaim, *The Cousinhood: The Anglo-Jewish Gentry* (London: Eyre and Spottiswoode, 1971).

Bernal, J.D., *The Freedom from Necessity* (London: Routledge and Kegan Paul Ltd, 1949).

——'The Material Theory of Life', *Labour Monthly* (review of *The Double Helix),* pp. 323–6, Jul. 1968.

——'Obituary notice of Rosalind Franklin', *Nature,* 182, p. 154, 19 Jul. 1958.

Boswell, J., *Life of Johnson* (London: George Routledge and Sons, 1895).

Bowles, John, *Viscount Samuel* (London: Gollancz, 1957).

Bragg, Melvyn, *On Giant's Shoulders* (London: Hodder and Stoughton, 1998).

——'The Dark Lady of DNA', *Independent on Sunday,* 15 Mar. 1998.

Brigstocke, Heather, 'Report by the High Mistress at Prizegiving, 27 Sep. 1983', *Paulina,* 1983.

Brook, Stephen, *The Club: The Jews of Modern Britain* (London: Constable, 1989; Pan, 1990).

Buchan, John, *The Thirty-Nine Steps* (London: Hodder and Stoughton, 1968).

Butler, P. Jonathan G., and Klug, A., 'The Assembly of a Virus', *Scientific American,* 239, 5, pp. 61–69, Nov. 1978.

Cairns, John, Stent, Gunther, and Watson, James D., *Phage and the Origins of Molecular Biology* (expanded edition: Cold Spring Harbor Press, 1992).

Calder, Ritchie, *Profile of Science* (London: George Allen and Unwin, 1951).

Carlisle, C.H., 'Serving My Time in Crystallography at Birkbeck: Some Memories Spanning Forty Years', unpublished; partly delivered as Valedictory Lecture, 30 May 1978.

Caspar, D.L.D., 'Structure of Tobacco Mosaic Virus', *Nature,* 177, pp. 928–30, 19 May 1956.

——and Klug, A., 'The Structure of Viruses as Determined by X-ray Diffraction', in C.S. Holton et al (eds), *Plant Pathology: Problems and Progress,* pp. 447–61 (Madison: University of Wisconsin Press, 1959).

Cavise, John, *The Genetic Gods: Evolution and Belief in Human Affairs* (Cambridge, Mass.: Harvard University Press, 1992).

——et al., 'The Composition of the Deoxypentose Nucleic Acids of Thymus and Spleen', *Journal of Biological Chemistry,* 177, p. 405, 1949.

Cesarani, David (ed.), *The Making of Modern Anglo-Jewry* (London: Blackwell, 1990).

——*The Jewish Chronicle and Anglo-Jewry, 1841–1991* (Cambridge: Cambridge University Press, 1994).

Chargaff, Erwin, 'Building the Tower of Babble', *Nature*, 248, pp. 776–9.

——*Essays on Nucleic Acids* (Amsterdam and London: Elsevier, 1963).

——'Quick Climb Up Mount Olympus', review of *The Double Helix* in *Science*, 159, pp. 1448–9, 1968.

——*Heraclitean Fire: Sketches from a Life before Nature* (New York: Rockefeller University Press, 1978)

Chomet, Seweryn, *Genesis of a Discovery* (London: Newman Hemisphere, 1995).

Cochran, W., Crick, F.H.C., and Vand, V., 'The Structure of Synthetic Polypeptides. I: The Transformation of Atoms on a Helix', *Acta Crystallographica*, 5, pp. 581–5, 1952.

Cooper, Artemis, *Writing at the Kitchen Table: Elizabeth David, The Authorized Biography* (London: Michael Joseph, 1999).

Cooper, Howard, and Morrison, Paul, *A Sense of Belonging: Dilemmas of British Jewish Identity* (London: Weidenfeld and Nicolson, with Channel 4 Television, 1991).

Cox, E.G., 'Summarised proceedings of a conference on the structures of semi-crystalline and non-crystalline materials', Nov. 1955, *British Journal of Applied Physics*, pp. 385–494, 7 Nov. 1956.

Crick, F.H.C., 'The Structure of the Hereditary Material', *Scientific American*, pp. 54–61, Oct. 1954.

——'Crick, Francis Harry Compton', *Les Prix Nobel en 1962*, pp. 69–70.

——'On the Genetic Code', Nobel Lecture, *Les Prix Nobel en 1962*, pp.179–87, 11 Dec. 1962.

——'How to Live with a Golden Helix', *The Sciences*, New York Academy of Sciences, pp. 6–9, Sep. 1979.

——*What Mad Pursuit? A Personal View of Scientific Discovery* (New York: Basic Books, 1988).

——and Watson, J.D., 'The Complementary Structure of Deoxyribonucleic Acid', *Proceedings of the Royal Society*, A 223, pp. 80–96, 1954.

Crook, E.M. (ed.), 'The Structure of Nucleic Acids and Their Role in Protein Synthesis', *Biochemical Society Symposia*, 14 (Cambridge: Cambridge University Press, 1957).

Cruickshank, D.W.D., 'Aspects of the History of the International Union of Crystallography', *Acta Cryst.*, A54, pp. 696–98, 1998.

Dainton, Frederick, and Thrust, B.A., 'Ronald George Wreyford

Norrish', *Biographical Memoirs of Fellows of the Royal Society*, 27, p. 406, 1981.

D'Arcy Hart, Ronald J. (ed.), *The Samuel Family of Liverpool and London: From 1755 Onwards* (London: Routledge and Kegan Paul, 1958).

Davies, K., *Cracking the Genome* (New York: Free Press, 2001).

Delbrück, M., 'On the Replication of Desoxyribonucleic Acid (DNA)', *Proceedings of the National Academy of Sciences*, 40, pp. 783–8, 1954.

Desiraju, Gautum R., 'The all-chemist', millennium essay, *Nature*, 408, p. 407, 23 Nov. 2000.

Dickens, Charles, *A Child's History of England* (London: Chapman and Hall, 1876).

Donohue, Jerry, 'Fragments of Chargaff', review of *Heraclitean Fire*, *Nature*, 276, pp. 133–5, 9 Nov. 1978.

——'Honest Jim?', *The Quarterly Review of Biology*, 51, pp. 285–89, Jun. 1976.

Dunitz, Jack D., 'Linus Carl Pauling', *The Royal Society*, pp. 317–38, 1996.

——'Recollections of Pre-Prebiotic Leslie Orgel', *Origins of Life and Evolution of the Biosphere*, 27, pp. 421–7, 1997.

Ellmann, Mary, *Thinking About Women* (New York: Harcourt Brace Jovanovich, 1968).

Ewald, P.P. (ed.), '50 Years of X-ray Diffraction', *International Union of Crystallography*, 1962.

Ferry, Georgina, *Dorothy Hodgkin: A Life* (London: Granta Books, 1998).

Fletcher-Jones, Pamela, *The Jews of Britain: A Thousand Years of History* (Moreton-in-the Marsh, Glos.: The Windrush Press, 1990).

Fowden, Leslie, and Pierpont, Stan, 'Norman Pirie (1907–97)', *Nature*, p. 560, 5 Jun. 1997.

Fraenkel-Conrat, H., and Singer, B., 'Virus reconstitution and the proof of the existence of genomic RNA', *Philosophical Transactions of the Royal Society*, B, 354, pp. 583–6, 1999.

Frank-Kamentskii, Maxim D. (trans. Lev Liapin), *Unravelling DNA: The Most Important Molecule of Life* (Reading, Mass.: Addison-Wesley, 1997).

Franklin, Arthur E., *Records of the Franklin Family and Collaterals* (London: Routledge, 1935).

——'The Attack on Mr Churchill', letter to *The Times*, 13 Dec. 1910.

Franklin, Colin, 'Vignettes of Rosalind', unpublished.

Franklin, Muriel, *Portrait of Ellis* (printed 'for private circulation' by Willmer Brothers Limited, London).

——'Rosalind', privately printed.

——'In the beginning', privately circulated memoir.

Franklin, Rosalind E., 'A note on the true density, chemical composition and structure of coals and carbonised coals', *Fuel*, xxvii–2, pp. 46–49.

——'A Study of the Fine Structure of Carbonaceous Solids by Measurements of True and Apparent Densities', Part I: Coals, *Trans. Faraday Soc.*, 45, pp. 274–86; and Part II: Carbonised Coals, pp. 668–82, both 1949.

——'A rapid approximate method for correcting low-angle scattering measurements for the influence of the finite height of the X-ray beam', *Acta Cryst.*, 3, pp. 158–9, 1950.

——'On the Influence of the Bonding Electrons on the Scattering of X-rays by Carbon', *Nature*, 14, pp. 71–2, Jan. 1950.

——'The Interpretation of Diffuse X-ray Diagrams of Carbon,' *Acta Cryst.*, 3, pp. 107–21, Jan. 1950.

——'Crystallite growth in graphitising and non-graphitising carbons', *Proc. Royal. Soc.*, A:209, pp. 196–218, 1951.

——'Structure of Graphitic Carbons', *Acta Cryst.*, 4, pp. 253–61, 1951.

——'Interim Annual Report: 1 Jan. 1951–1 Jan. 1952'; 'Interim Annual Report: Jan. 1952-Jan. 1953'; 'Annual Report: 1 Jan. 1953–1 Jan. 1954, Crystallographic Laboratory, Birkbeck College, 21 Torrington Square, WC1.

Franklin, R., and Gosling, R., 'Molecular Configuration in Sodium Thymonucleate', *Nature*, pp. 740–1, 25 Apr. 1953.

——'Evidence for 2-Chain Helix in Crystalline Structure of Sodium Deoxyribonucleate', *Nature*, pp. 156–7, 25 Jul. 1953.

——'The structure of Sodium Thymonucleate Fibres: I. The Influence of Water Content', 'The Structure of Sodium Thymonucleate Fibres: II. The Cylindrically Symmetrical Patterson Function', *Acta Cryst.*, 6, pp. 673–7 and 678–85, 1953.

——'The Structure of Sodium Thymonucleate Fibres: III. The Three-Dimensional Patterson Function', *Acta Cryst.*, 8, 1955.

——'Le Rôle de l'Eau dans la Structure de l'Acide Graphitique', *Journal de Chemique et Physique*, 50C, p. 26, 1953.

——'Structure of tobacco mosaic virus', *Nature*, 175, p. 379, 26 Feb. 1955.

——'Structural resemblances between Schramm's repolymerised

A-protein and tobacco mosaic virus', *Biochimica et Biophysica Acta,* 18, p. 313, 1955.

Franklin, Rosalind E., and Klug, A., 'The splitting of the layer lines in the X-ray fibre diagrams of helical structures: Application to tobacco mosaic virus', *Acta Cryst.*, 1, p. 777, 1955.

——'The nature of the helical groove on the tobacco mosaic virus particle', *Bioc. et Biophys. Acta,* 19, p. 403, 1956.

Franklin, Rosalind E., and Commoner, B., 'X-ray diffraction by an abnormal protein (B8) associated with tobacco mosaic virus', *Nature,* 175, p. 1076, 18 Jun. 1955.

——'Progress Report of the Agricultural Research Council research group in Birkbeck College Crystallography Laboratory, for the year 1955.'

——'Location of the ribonucleic acid in the tobacco mosaic virus particle', *Nature,* 177, p. 928, 1956.

——'X-ray diffraction studies of Cucumber Virus 4 and three strains of tobacco mosaic virus', *Bioc. et Biophys. Acta,* 19, p. 403, 1956.

Franklin, Rosalind E., and Holmes, K.C., 'The helical arrangement of the protein subunits in tobacco mosaic virus', *Bioc. et Biophys. Acta,* 21, p. 405, 1956.

——'Tobacco Mosaic Virus: Application of the Method of Isomorphous Replacement to the Determination of the Helical Parameters and Radial Density Distribution', *Acta Cryst.*, 11, pp. 213–20, 1958.

Franklin, Rosalind E., Klug, A., and Holmes, K.C., 'X-ray Diffraction Studies of the Structure and Morphology of Tobacco Mosaic Virus', *Ciba Foundation Symposium on The Nature of Viruses,* pp. 39–52 (London: J.&.A. Churchill, 1956).

——'Homogeneous and Heterogeneous Graphitisation of Carbon', *Nature,* 177, p. 239, 4 Feb. 1956.

——'Structure of Tobacco Mosaic Virus: Location of the Ribonucleic Acid in the Tobacco Mosaic Virus Particle' (two articles), *Nature,* 177, pp. 928–30, 19 May 1956.

——'X-ray diffraction studies of the structure of the protein in tobacco mosaic virus', in Neuberger, A. (ed.), *Symposium on Protein Structure* (London: Methuen, 1958).

——'On the Structure of Some Ribonucleoprotein Particles', *Trans. Faraday. Soc.*, 25, p. 197, 1958.

Franklin, R.E., Klug, A., Finch, J.T., and Holmes, K.C., 'On the Structure of Some Ribonucleoprotein Particles', *Discussions of the Faraday Society,* No. 25, 197–8, 1958.

Franklin, (The Late) R.E., and Klug, A., 'The Structure of RNA in Tobacco Mosaic Virus and in some other Ribonucleoproteins', *Trans Faraday Soc*, 435, Vol. 55, Part 3, pp. 494–5, Mar. 1959.

Fraser, Mary J., and Robert, D.B., 'Evidence on the Structure of Deoxyribonucleic Acid from Measurements with Polarised Infra-Red Radiation', *Nature*, pp. 761–2, May 1951.

Fraser, Dr. R.D.B., 'The Structure of Deoxyribose Nucleic Acid', written at M. Wilkins's request on 17 Mar. 1953, unpublished.

Friedman, Meyer, and Friedland, Gerald W., *Medicine's 10 Greatest Discoveries* (New Haven and London: Yale University Press, 1998).

Furberg, Sven, 'On the Structure of Nucleic Acids,' *Acta Chemica Scandinavica*, 6, pp. 634–40, 1952.

——'My Work on Nucleic Acid Structure 1947–49', Institute of Chemistry, University of Oslo.

Furth, R., 'The Physics Department of Birkbeck College', The Institute of Physics, *Bulletin*, 55, pp. 150–8, Jul. 1954.

Gilmer, Sander L., *Jewish Self-Hatred: Anti-Semitism and the Hidden Language of the Jews* (Baltimore and London: Johns Hopkins Press, 1986).

Glynn, Jenifer, *Tidings from Zion* (London: I.B. Tauris and Co. Ltd, 2000).

——'Rosalind Franklin 1920–1958', in Blacker, Carmen, and Shils, Edward (eds), *Cambridge Women: Twelve Portraits* (Cambridge: Cambridge University Press, 1996).

Goertzel, Ted, and Goertzel, Ben, *Linus Pauling: A Life in Science and Politics* (New York: Basic Books, 1995).

Goldsmith, Maurice, *Sage: A life of J.D. Bernal* (London: Hutchinson, 1980).

Goldstein, David, Jacobs, Louis, and Lipman, Vivian D. (eds), *The Century of Moses Montefiore* (Oxford: Oxford University Press, 1988).

Gosling, R.G., 'X-ray Diffraction Studies of Desoxyribose Nucleic Acid', thesis submitted for the Degree of Doctor of Philosophy in the University of London, Feb. 1954.

——'Abstract of Thesis submitted by R.G. Gosling for the PhD Examination'.

Gratzer, Walter, 'The hunting of the helix', review of the reissue of *The Double Helix* and *The Eighth Day of Creation*, *Nature*, pp. 344–5, 27 Mar. 1997.

——*Eurekas and Euphorias: The Oxford Book of Scientific Anecdotes*, (Oxford: Oxford University Press, 2002).

Hager, Tom, *Force of Nature: Life of Linus Pauling* (New York: Simon and Schuster, 1996).

Hamilton, L.D., 'DNA: Modes and Reality', *Nature*, 18 May 1968.

Hamminga, Harmke, 'The International Union of Crystallography: Its Formation and Early Development', *Acta Cryst.*, A 45, pp. 581–601.

Hargittai, Istvan, 'Aaron Klug', 'James Watson', and 'Erwin Chargaff' in *The Chemical Intelligencer*, pp. 4–13 and pp. 20–24, Oct. 2000; pp. 4–9, Jan. 1998.

Hargittai, Istvan, and Hargittai, Magdolna, *In Our Own Image: Personal Symmetry in Discovery*, (New York: Kluwer Academic, 2000).

Harris, Peter, 'On charcoal', *Interdisciplinary Science Reviews*, 24, 4, pp. 301–6, 1999.

——'Rosalind Franklin's work on coal, carbon, and graphite', *Interdisciplinary Science Reviews*, 26, 3, pp. 204–9, 2001.

Harrison, B.D., and Wilson, T.M.A. (eds.), 'Tobacco mosaic virus: pioneering research for a century', *Philosophical Transactions of the Royal Society*, 354, 1383, pp. 517–685, Mar. 1999.

Harrison, J.F.C., *A History of the Working Men's College 1854–1954* (London: Routledge and Kegan Paul, 1954).

Holmes, K.C., and Franklin, Rosalind E., 'The Radial Density Distribution in Some Strains of Tobacco Mosaic Virus', *Virology*, 6, pp. 328–36, 1958.

Hubbard, Ruth, 'The Double Helix: A Study of Science in Context', *The Politics of Woman's Biology* (New Brunswick, N.J.: Rutgers University Press, 1990).

——'The Story of DNA', in Senechal, Marjorie (ed.), *Structures of Matter and Patterns in Science* (Cambridge, Mass.: Schenkman Publishing Company Inc., 1980).

Hussain, Farooq, 'Did Rosalind Franklin deserve DNA Nobel prize?', *New Scientist*, p. 470, 20 Nov. 1975.

Inagaki, Michio, 'Rosalind E. Franklin – Who was She?', *Energeia*, University of Kentucky, Center for Applied Energy Research, Vol. 6, No. 6, p. 1, pp. 3–4, 1995.

Inwood, Stephen, *A History of London* (London: Macmillan, 1998).

Jacob, François, *The Statue Within: An Autobiography* (London: Unwin Hyman).

Janeway, Elizabeth, *Man's World, Woman's Place: A Study in Social Mythology* (London: Michael Joseph, 1971).

Jevons, Fred, and Stokes, Terry, *Winner Take All: Case Study of the*

Double Helix (Victoria, Australia: Deak University Press, 1979).

Judson, Horace Freeland, *The Eighth Day of Creation: Makers of the Revolution in Biology* (London: Jonathan Cape, 1979, Penguin, 1995; expanded edition: New York: Cold Spring Harbor Press, 1996).

——'Reflections on the Historiography of Molecular Biology', *Minerva*, Vol. XVIII, 3, pp. 369–421, Autumn 1980.

——'Reweaving the Web of Discovery: Why are really first-rate biographies of scientists so rare?', *The Sciences*, Nov.–Dec. 1983, pp. 44–53.

Julian, Maureen M., 'Women in Crystallography', in Kass-Simon, G., and Farnes, P. (eds), *Women of Science: Righting the Record* (Bloomington: University of Indiana Press, 1993).

Julius, Anthony, *T.S. Eliot, Anti-Semitism and Literary Form* (Cambridge: Cambridge University Press, 1995).

——'England's Gifts to Jew Hatred', *Spectator*, 11 Nov. 2000.

Kamminga, Harmke, 'The International Union of Crystallography: Its Formation and Early Development', *Acta Cryst.*, A 45, pp. 581–601, 1989.

Keene, Joane (ed.), 'Norland Place School' (Richmond, Surrey: E.H. Baker and Co. Ltd, 1976).

Kendrew, John C., 'How molecular biology started,' review of *Phage and the Origins of Molecular Biology* in *Scientific American*, 216, 3, pp. 343–4, 1967.

Kerslake, Jean, 'Memories of Rosalind 1933–1938,' privately circulated booklet.

Kime Scott, Bonnie (ed.), *Selected Letters of Rebecca West* (New Haven: Yale University Press, 2000).

Klug, A., 'From Macromolecules to Biological Assemblies', Nobel lecture, 8 Dec. 1982.

——'Commentary on the Two Papers by Caspar and Franklin', in Scholthof, K.B.G., Shaw, J.G., and Zaitlin, M. (eds), *Tobacco Mosaic Virus: One Hundred Years of Contribution to Virology* (St Paul, Minnesota: American Phytopathology Society Press, 1999).

——'Rosalind Franklin and the Discovery of the Structure of DNA', *Nature*, Vol. 219, 24 Aug. 1968, pp. 808–10 and 843–4. Also corrigenda pp. 879, 1192 and correspondence p. 880. Corrigendum. *Nature*, Vol. 219, 14 Sep. 1968, p. 1192.

——'Rosalind Franklin and the Double Helix', *Nature*, Vol. 248, 26 Apr. 1974, p. 78.

Klug, A., and Caspar, D.L.D., 'The Structure of Small Viruses' (dedicated to the memory of Rosalind Franklin), *Advances in Virus Research*, 1960.

——'The tobacco mosaic virus particle: structure and assembly', *Phil. Trans. Royal Soc.*, 354, pp. 531–5, 1999.

Klug, A., Finch, J.T., and Franklin, R.E., 'Structure of turnip yellow mosaic virus', *Nature*, 170, p. 683, 1957.

——'Structure of turnip yellow mosaic virus: X-ray diffraction studies', *Biochim. et Biophys Acta*, 25, p. 242, 1957.

Klug, A. and Franklin, R.E., 'The reaggregation of the A-protein of tobacco mosaic virus', *Biochim. et Biophys Acta*, 23, p. 199, 1957.

——'Order-disorder transitions in the structures containing helical molecules', *Trans. Faraday Soc.*, 25, p. 104, 1958.

Klug, A., Franklin, R.E. and Humphreys-Owen, S.P.F., 'The crystal structure of Tipula iridescent virus as determined by Bragg reflection of visible light' *Biochim. et Biophys. Acta*, 32, p. 203, 1959.

Lipman, V.D., *A History of the Jews in Britain Since 1858* (Leicester: Leicester University Press, 1990).

Lonsdale, Kathleen, 'Women in science: Reminiscences and reflections', *Impact of Science on Society*, Vol. XX, No. 1, pp. 45–59, 1970.

Mackay, A.L., 'Recent Soviet Work in the Field of Crystallography' and 'Crystallography in Eastern Europe', *Supplemento al Volume X. Serie IX Del Nuovo Cimento*, 4, pp. 387–414, 1953.

Maddox, Brenda, 'Franklin Recalled', *Nature*, 398.

——'The Dark Lady of DNA', *Observer*, 6 Mar. 2000.

Maddox, John, 'The architecture of viruses', *Manchester Guardian*, 30 Jun. 1959.

Margerison, Tom, 'The Architects of Life', *Sunday Times*, 9 Dec. 1962.

Mason, Joan, 'The Admission of the First Women to The Royal Society of London', *Notes Rec. Royal Soc. London*, 46(2), pp. 279–300, 1992.

——'The Women Fellows' Jubilee', *Notes Rec. Royal Soc. London*, 49 (1), pp. 125–40, 1995.

Matriculation and School Examinations Council, 'Report of the Council on the Educational Work of St Paul's School for Girls', University of London, Feb. 1935.

May, Ernest R., *Strange Victory: Hitler's Conquest of France* (London: I.B. Tauris and Co. Ltd, 2000).

Meitner, Lise, 'The Status of Women in the Professions', *Physics Today*, pp. 16–21, Aug. 1960.

Merkin, Daphne, 'Growing Up Rich', *New Yorker*, 26 Apr. and 3 May 1999.

Meurig-Thomas, John, 'Molecular Biology and Peterhouse', *Chemical Intelligencer*, pp. 25–33, Oct. 2000.

Moorehead, Caroline, 'What the right parents can do for a girl: Dorothy Hodgkin: Nobel Prize chemist', *The Times*, 21 Apr. 1975.

Neilands, Robin, *The Bomber War: Arthur Harris and the Allied Bomber Offensive 1939–1945* (London: John Murray, 2001).

Nicholson, W., *Life Story*, Horizon, BBC2, 27 Apr. 1987.

Ogilvie, Marilyn Bailey, with Meek, Kerry Lynne (eds), *Women and Science, An Annotated Bibliography* (New York and London: Garland, 1996).

Olby, R., 'Before *The Double Helix*', *New Scientist*, 27 Jun. 1968.

——'Francis Crick, DNA, and the Central Dogma', *Daedalus*, XCIX, 4 (Fall 1970).

——'Rosalind Elsie Franklin' in Gillespie, Charles C. (ed.), *Dictionary of Scientific Biography* (New York: Charles Scribner's Sons, 1972).

——*The Path to the Double Helix* (London: Macmillan, 1974).

Opfell, Olga S., *The Lady Laureates: Women Who Have Won the Nobel Prize* (Metuchen, N.J. and London: The Scarecrow Press, 1986).

Ormrod, Sarah J. (ed.), *Cambridge Contributions*, especially Sutherland, Gillian, 'Nasty Forward Minxes', pp. 88–102 (Cambridge: Cambridge University Press, 1998).

Overbye, Dennis, *Einstein in Love* (London: Bloomsbury, 2001).

Pauling, Linus, and Corey, Robert B., 'A Proposed Structure for the Nucleic Acids', *Proc. Nat. Acad. Sciences*, 39, 2, pp. 84–97.

——'Compound Helical Configuration of Polypeptide Chains', *Nature*, 171, 59, Feb. 1953.

Pauling, Peter, 'DNA – the race that never was?', *New Scientist*, pp. 558–60, 31 May 1973.

Pert, Candace B., *Molecules of Emotion: Why You Feel the Way You Feel* (New York: Simon and Schuster, 1997).

Perutz, Max F., 'Co-chairman's remarks: before the double helix', *Gene*, 135, 1993.

——'Gene genius', *Times Higher Education Supplement*, Millennium Edition, 31 Dec. 1999.

——'How the secret of life itself was discovered', *Daily Telegraph*, 27 Apr. 1987.

——letter to *Science*, 164, 1969.

Pierpont, W. S., 'A very sceptical biochemist: Norman Wingate "Bill" Pirie FRS (1907–1997)', *The Biochemist*, Aug. 1997.

——'Norman Wingate Pirie', *Biographical Memoirs of Fellows of the Royal Society*, pp. 399–415, 1999.

——'Norman Pirie', *Independent*, 22 Apr. 1997.

Piper, Anne, 'Light on a dark lady', *Trends in Biochemical Science*, 23 Apr. 1998.

Quinn, Susan, *Marie Curie: A Life* (New York: Simon and Schuster, 1995).

Randall, J.T., 'Dr. H.A.H. Boot and the cavity magnetron', *New Scientist*, pp. 602–3, 3 Mar. 1983.

——'Notes on Current Research prepared for the visit of the Biophysics Research Committee, 15 Dec. 1952', Biophysics Research Unit, Wheatstone Physics Laboratory, King's College London, 5 Dec. 1952, MRC. 52/815; PRO/FD 1/1702 66670.

——'The Research Work of the Wheatstone Physics Laboratory, King's College, University of London', *British Science News*, 2, 14, pp. 43–7.

——'Emmeline Jean Hanson', *Biographical Memoirs of Fellows of the Royal Society*, 21, pp. 313–44, Nov. 1975.

——'An Experiment in Biophysics', *Proc. Royal Soc.*, 208, 1, Mar. 1951.

Rich, Alexander, 'The Nucleic Acids: A Backward Glance', reprinted from *DNA: The Double Helix, Annals of the New York Academy of Sciences*, Vol. 758, 30 Jun. 1995.

Ricks, Christopher, *T.S. Eliot and Prejudice* (London: Faber, 1988).

Samuel, The Rt Hon. Viscount Samuel, 'The Future of Palestine', CAB 37/123/43, 21 Jan. 1915.

——*Memoirs* (London: Cresset Press, 1945).

——*Creative Man and Other Addresses* (London: Cresset Press, 1949).

Sardan, Ziauddin, 'A male preserve', *New Statesman*, 13 Nov. 1999.

Sayre, Anne, *Rosalind Franklin & DNA* (New York: Norton, 1975).

——'The case for Rosalind Franklin', *Sunday Times*, 4 Apr. 1976.

Schiebinger, Londa, *Has Feminism Changed Science?* (Cambridge, Mass.: Harvard University Press, 1999).

Schlesinger Jr, Arthur M., *A Life in the 20th Century: Innocent Beginnings, 1917–1950*, (Boston: Houghton Mifflin, 2000).

Schrödinger, Erwin, *What Is Life?* (Cambridge: Cambridge University Press, 1992 edition).

Segev, Tom, *One Palestine, Complete: Under the British Mandate* (London: Little, Brown, 2000).

Serafini, Anthony, *Linus Pauling: A Man and His Science* (New York: Simon and Schuster, 1989).

Shearer, Benjamin F., and Shearer, Barbara S., 'Rosalind Elsie Franklin (1920–1958)' in *Notable Women in the Life Sciences: A Biographical Dictionary* (Westport, Conn. and London: Greenwood Press, 1996).

Sime, Ruth Lewin, *Lise Meitner: A Life in Physics* (Berkeley and Los Angeles: University of California Press, 1996).

Slyser, Mels, '*Waarom kreeg "Rosy" geen Nobelprijs?*', *Vrij Nederland*, pp. 50–51, 15 Aug. 1998.

Snow, C.P., *The Search* (London: Macmillan, 1958).

Sofer, Sasson, (trans. Dorothea Shefet-Vanson), *Zionism and the Foundations of Israeli Diplomacy* (Cambridge: Cambridge University Press, 1998).

Sonnert, Gerhard, with Holton, Gerald *Gender Difference in Scientific Careers* (New Brunswick, N.J.: Rutgers University Press, 1995).

Swann, Brenda, and Aprahamian, Francis (eds.), *J.D. Bernal: A Life in Science and Politics*, including Trent, Peter, 'Bernal the Scientist', and Preface by Eric Hobsbawm (London and New York: Verso, 1999).

Sykes, Bryan, *The Seven Daughters of Eve* (London: Bantam, 2001).

Tift, Susan E., and Jones, Alex S., '*The Times* and the Jews', *New Yorker*, 19 Apr. 1999.

Tucker, Anthony, 'Leaves on the global menu', obituary of N.W. Pirie, *Guardian*, 31 Mar. 1997.

Wasserstein, Bernard, *Herbert Samuel: A Political Life* (Oxford: Oxford University Press, 1992).

Watson, James D., 'The Structure of Tobacco Mosaic Virus: I. X-ray Evidence of a Helical Arrangement of Sub-units Around the Longitudinal Axis', *Bioc. et Biophys. Acta*, 13, pp. 10–19, 1954.

——'The Involvement of RNA in the Synthesis of Proteins', Nobel Lecture, 11 Dec. 1962.

——'James Dewey Watson', *Les Prix Nobel en 1962*, (Stockholm: Almquist and Wiksell International, 1963).

——reply to Nobel Prize awards, Stockholm, Dec. 1962, on behalf of all three recipients, *Les Prix Nobel en 1962*.

——*The Double Helix: A Personal Account of the Discovery of the Structure of DNA* (New York: Athenaeum, 1968; London: Weidenfeld and Nicolson, 1981).

——*The Double Helix*, Norton Critical Edition, ed. Gunther S. Stent, 1980 (New York and London: W.W. Norton and Company, 1980).
——*Genes, Girls and Gamow* (Oxford: Oxford University Press, 2001).
——*A Passion for DNA: Genes, Genomes, and Society*, with an introduction, afterword and annotations by Walter Gratzer (New York: Cold Spring Harbor Press, 2000; Oxford: Oxford University Press, 2000).
——*Times Love, Memory: A Great Biologist and His Quest for the Origins of Behaviour* (New York: Knopf, 1999).
Watson, J.D., and Crick, F.H.C., 'A Structure for Deoxyribonucleic Acid', *Nature*, 171, pp. 737–8, 25 Apr. 1953.
——'Genetical Implications of the Structure of Deoxyribonucleic Acid', *Nature*, p. 964, 30 May 1953.
——'The Structure of DNA', *Cold Spring Harbor Symposia on Quantitative Biology*, 18, 123, 1953.
Watt, J.D., and Franklin, Rosalind E., 'Change in the Structure of Carbon during Oxidation', *Nature*, 180, pp. 1190–91, 30 Nov. 1957.
Waugh, Evelyn, *Vile Bodies* (London: Methuen, 1978).
Weill, Adrienne R., '*Etude aux rayons X de la fragilité de revenu d'un acier à faibles teneurs en nickel et en chrome*', *Comptes Rendus Académie des Sciences*, Séance du 13 février 1950.
White, Michael, *Rivals: Conflict as the Fuel of Science* (London: Secker and Warburg, 2001).
Wilkins, Maurice, 'Wilkins, Maurice Hugh Frederick', *Les Prix Nobel en 1962*, pp. 74–5.
——'John Turton Randall', *Biographical Memoirs of Fellows of the Royal Society*, Vol. 33, 1987.
——'I. Ultraviolet Dichroism and Molecular Structure in Living Cells. II. Electron Microscopy of Nuclear Membranes', *Estratto dalle Pubbl. Staz. Zool. Napoli.*, Vol. 23, Supplemento, pp. 105–15.
Wilkins, Maurice, Stokes, A.F., and Wilson, H.R., 'Molecular Structure of Deoxypentose Nucleic Acids', *Nature*, pp. 738–40, 25 Apr. 1953.
——'The Molecular Configuration of Nucleic Acids', *Les Prix Nobel en 1962* (Stockholm, 1963).
——'Acid: an Extensible Molecule?', 'Physical Studies of Nucleic Acid', *Nature*, Vol. 167, No. 4254, pp. 759–60, 12 May 1951.
Wilkins, Maurice, and Randall, J.T., *Bioc. et Biophys. Acta*, 10, p. 192, 1953.

Wilkins, Maurice, Zuaby, G., and Wilson, H.R., 'X-ray Diffraction Studies of the Structure of Deoxyribonucleoprotein', *Transactions of the Faraday Society*, No. 435, Vol. 55, Part 3, Mar. 1959.

Wilson, Edward O., *Consilience* (New York: Knopf, 1998).

Wilson, H.R., 'The double helix and all that', *Trends in Biochemical Sciences*, 13, 7, pp. 275–8, Jul. 1988.

——'Connections', *TIBS*, 26, 5, pp. 334–7, May 2001.

INDEX